D0502994

MEN AND DISCOVERIES IN MATHEMATICS

IN THE SAME SERIES

Men and Discoveries in Electricity

Men and Discoveries in Chemistry

Men and Discoveries in Mathematics

BRYAN MORGAN

JOHN MURRAY

Printed in Great Britain by
Cox & Wyman Ltd
London, Fakenham and Reading
0 7195 2587 x

To
V and MD

Contents

Illustrations

between pp. 104/5

Reproduced by permission of the Science Museum – Crown Copyright

Acknowledgements and Biblography

The author would like to thank all who have helped with the compilation of this book (particularly Messrs J. Coop and J. Dewhurst and the staff of the Science Museum Library) and with its checking. Like its companions, it has also benefited in planning and production from the understanding help of Mr John Murray and Mrs Osyth Leeston. Such outside advice, however, does not relieve a writer from the need to stress that his errors – and his opinions – are his own.

The line drawings were sketched by the author and prepared for press by Eagle Arts. Photographic credits for the plates will be found on the previous page.

Like its predecessors, too, this book has been designed to present not new facts but available facts in a new way; and it has hence drawn on those histories of mathematics which themselves reach back certainly for 400 years and perhaps to Greek times. Although Moritz Cantor's four volumes in German form the definitive work down to 1800, the standard source in English is that of F. Cajori, which also adds most of another century to the story. This was last published more than fifty years ago, however, as was W. W. Rouse Ball's shorter but more readable (and perhaps reliable) classic. A recent two-volume history – to which, despite its curious organisation, the present author is much indebted – is that of the American D. E. Smith: more recent still, but selective, ending around 1800 and needing for full appreciation some algebraic background, is J. F. Scott's single volume. L. Hogben's *Mathematics in the Making* is opinionated and rather hard going, but contains some good expositions of mathematical ideas.

There are also, of course, a number of specialised histories of particular branches of mathematics.

Of the more biographical works, T. E. Bell's *Men of Mathematics* is probably the most scholarly and certainly the most individualistic: it contains little on the period before 1600. Also worth attention are A. Hooper's *Makers of Mathematics* (an American-based book, presumably for children, which is generally well-written but at some times over-detailed and at others misleading) and H. W. Turnbull's very concise *The Great Mathematicians*. An interesting but somewhat dated and inaccessible miscellany is A. Macfarlane's *Ten British Mathematicians*: G. Prasad's *Some Great Mathematicians of the Nineteenth Century* is of the same type.

The few English-language biographies of individual mathematicians are also rather dated and hard to find, typical works of this type being T. L. Heath on the Greeks, E. S. Haldane on Descartes and D. Brewster on Newton – for whom there is also an easily-read modern life by E. M. Andrade. Finally, H. Midonick's *The Treasury of Mathematics* should be recommended to those interested in studying excerpts from original papers.

London 1972 B. M.

Foreword

WHAT THIS BOOK IS ABOUT

One man presses a button to launch a space-craft: another works on the design for a motorway bridge: and a third is engaged on a market survey for a new product. Most people would recognise all these activities as examples of applied mathematics, and be aware that they involve a kind of thinking which lies behind a hundred other aspects of modern life. They are also, perhaps, conscious that there is no division in *kind* – but only in *purpose* – between such mathematics-in-action and the 'pure' study of similar triangles and simultaneous equations which is at the centre of most school courses.

Both types of mathematics, too, have been worked out by the same kind of man: indeed, the very greatest names in the subject, such as those of Archimedes, Newton and Gauss, shine equally brightly in both its aspects. And yet the mathematician himself remains for most of us a shadowy figure; we may no longer picture him as a shaggy-haired eccentric, but we cannot summon-up (as we can for the scientist) a stereotype of a man in a white coat with a slide-rule. The first aim of these pages, then, is to present the great men of mathematics – men whose lives are linked in an almost-unbroken chain extending over more than 2,500 years – in a way which will show how their achievements, their personalities and the society in which they worked were intertwined throughout one of the most exciting and continuing of all the adventures of the human mind.

This book is the third in a series of which the other two are devoted to experimental sciences; and though in some ways it differs from its predecessors, it has like them been written for a particular readership and in a particular form. The audience envisaged, for instance, is again one made up largely of students – at school or in college, in Britain or overseas, and including those who are not themselves reading technical subjects but who believe that an all-round education must include some understanding of that revolution of knowledge which has done more than any political changes to shape the world we live in. But once more, too, it is hoped that older readers as well will appreciate a compact and self-contained account of some of the ideas and discoveries which are woven together in today's fabric of scientific thinking.

Every discipline of thought has its own historical shape, though, and it was never conceived that these books should follow a rigid formula. In the present case, for example, those fundamental ideas which form the raw material of mathematics – but which are less familiar and less appreciated by the non-specialist than are magents and acids – demand so much space that the strictly biographical content of these pages must be correspondingly less. Furthermore, we shall still only be selecting themes out of many possibilities; for whereas earlier books could claim to introduce almost all the key men and ideas in their respective subjects, we must here choose from a much wider range of important material.

Some readers may even wonder why mathematics has been included in the series, for though this hard-to-define discipline has been termed the queen of sciences its nature is in reality rather more complex. Mathematics has also been called a *servant* of science; and it is certainly a major tool of some sciences, just as logic and the experimental philosophy are tools of them all. But (contrary to most people's first ideas) its usefulness turns out to be very unevenly distributed.

Thus, mathematics plays almost no part in 'pure' biology, its main contribution being *via* its specialised branch of statistics whose methods are just as applicable to a host of non-scientific studies. It is equally remote from the centres of geology and other 'observational' sciences. In chemistry it is called in mainly on the fringes of research, and indeed the development of that subject was actually delayed by its attempt to use mathematical methods before it was ready for them. Only physics – that science with a dozen major divisions which has also absorbed much of astronomy – has since its systematic beginnings nearly 400 years ago regularly depended upon mathematics to interpret its experimental results and to suggest new theories, though it has done this so completely that the deepest ideas in modern physics can barely be translated into words.

The historic importance of mathematics in technology and the applied sciences, too, is easy to over-rate, for perhaps only in navigation has it ever played a decisive role. From the age of the lever and the screw, on through those of the steam engine and the telephone, and so up to the present era of the transistor and the colour TV set, technical advances have usually come about through a mixture of insight, perseverance, inspired commonsense, experience and sheer luck, with that systematic analysis which is the concern of applied mathematics playing only the smallest of parts. Even in engineering, mankind progressed from the epoch of felled tree-trunk bridges to that of the great iron spans of the early railway age mainly 'by guess and by God', with mathematicians adding little to design until well after 1850. It is normally only *after* a man has made a sailing-boat, a steam engine or a cyclotron that the mathematical physicist comes along, tells him just how the machine works (which may be rather different from the inventor's own idea), and add: 'But, surely, it would be better if you did it like this . . .'

We should also note here that mathematics differs from the experimental sciences in its philosophy: for whereas the latter are based on both the *inductive* reasoning which leads from a dozen particular instances up to a general rule and the *deductive* type which leads from that rule back to predicting further instances, only the second type of argument is permitted in formal mathematics. This does not mean that the mathematician necessarily *works* logically, for he is as much dependent on 'hunches' and 'thinking sideways' as is the scientist. But his publications must be based on pure analysis and not behaviour, and indeed the mathematician *as a mathematician* is not very interested in whether his results arise from, or even accord with, experience. All that matters to him is their self-consistency and freedom from contradictions.

Despite all these reservations, the contribution of mathematics to physics in its widest sense – to man's knowledge of the working of the machinery of the universe, and hence to his control over his world – would alone be enough to demand the inclusion of a book on its history in this series. It should also be remembered that the usefulness of mathematics (and the word itself, first used some 2,500 years ago, means simply 'method') is not confined to subjects usually thought of as being technical, for real mathematics – as opposed to simple figuring and gadgetry – has played a part in warfare since Plato thought that a general ought to understand the theory of numbers so as to be able to divide up his troops into squares and triangles, in finance for some 750 years, and at the gambling tables for over three centuries. Today it is to be found at work in almost every form of the 'useful arts'.

There is, indeed, a school of thought which holds that the ultimate justification of mathematical thought is to be found in such applications. Even upholders of this minority view, though, would admit that in mathematics – as in the sciences – progress in the applied branches has time and again been accelerated (or even made possible) only as a result of work earlier carried out from pure intellectual curiosity. The astounding usefulness of the useless is, in fact, a moral which will emerge from almost every page of this book.

Those interested in following up a controversy as old as the biblical conflict between Jubal and Tubal Cain will find spokesmen for the contrasting views in books by two British academics, with Lancelot Hogben defending mathematics for use and G. H. Hardy claiming that its nature is a very different one and that mathematics can be justified only as an adventure of the human mind. Whether that nature is itself closer to that of a game or an art is another matter, for the question 'What is mathematics?' has never been satisfactorily answered. But the subject certainly exists at one remove from what the world thinks of as 'reality'.

This 'otherness' of mathematics is indeed a further theme which will run throughout this book: in the extreme case, it has been said, the mathematician is a man who carries out operations which he cannot explain upon undefinable objects. It is interesting here to compare two quotations, of which one runs '[This] is the subject in which we never know what we are talking about nor whether what we are saying is true' and the other '[In this world] all facts and all beliefs cease to be true or false and become interesting possibilities.' The first remark was made by Bertrand Russell about mathematics, and the second by W. H. Auden when exploring the nature of poetry.

Even granted that 'pure' mathematics define an abstract world whose links with the visible one seem almost accidental, a trap into which non-mathematicians since Aristotle have often fallen is to suppose that they are primarily concerned with the handling of *numbers*. On the contrary, there are whole fields of mathematics – including those which have proved of the greatest interest in the last hundred years – into which specific numbers hardly enter at all. And some of these are even independent of the general *idea* of number.

Those educationists who believe that the world will never bridge the gap between the 'two cultures' of the arts and the sciences until mathematicians become more literate and non-mathematicians more 'numerate' have hence confused an issue as well as coined an unnecessary and ugly word; for the mathematician's own mind is not so much 'numerate' as 'symbolate', and he is probably so bad at simple addition that he leaves the household accounts to his wife. When he writes 'X' he may be using that mark to represent a number, a group of related numbers, or something to which a number can meaningfully be attached such as a length or a likelihood. But his 'X' is just as likely to be standing for a logical proposition, a pattern, an operation, or even a concept which cannot be translated into any words, let alone numbers.

A glance at a typical modern text-book shows how almost alarmingly independent the pure mathematician is of words and their meanings – as independent as the musician whom he in some ways resembles. He may well begin with the phrase 'Consider the function . . .' and then in forty pages have no more use for English than an occasional preposition. But *this* book is written in words; and it must be confessed that, though many individual mathematicians have been talented writers, only psychology has tortured those other tools of thought further from a rational usage than has mathematics.

It is indeed ironic that what claims to be the most rigidly logical of all disciplines should have mis-handled its vocabulary in almost every way possible. Some of its familiar terms, such as *sine*, are due to bad translations: it uses several words (*power*, *index*, *exponent*) for almost the same idea and the same words (*base*, *root*, *radix*) for several different ideas: it takes terms from common parlance (*group*, *real*, even *space*) and gives them highly-specialised and artificial meanings: and it christens its various branches (*topology*, and the indefensible '*analysis*') according to the whims of fashion rather than on a plan bearing any relation to reason and the meaning of language. The author has tried here to employ mathematical terms in the way which will cause the reader least confusion, and sees no need to apologise if he has occasionally used words such as 'function' or 'equation' in a sense subtly different from that currently approved by the professionals working in a specialised field.

Symbolism, too, is very far from being standardised on even the simplest level. But here the mathematician has to shape his usage to the few forms available to the ordinary printer, and so has a good excuse for the fact that even such devices as x and . have quite different meanings in different branches of his subject. It should in the present book be clear from the context that (for instance) AB in a geometrical discussion implies the distance between two points and in an algebraic one the product of two quantities. To ask 'What does AB mean in mathematics?' is rather like asking 'Do the laws of football allow a player to handle the ball?' without specifying whether the game is rugger, soccer or some other variety. All that matters, in mathematics as in football, is that one cannot change the rules at half-time.

This mention of mathematical language itself leads on to what is perhaps the most important question to be asked before we can begin our journey through time, which concerns how much the reader is expected to know of the subject at the outset. It would be pleasant to answer 'Nothing at all'. But whereas some people may have 'missed out' completely on chemistry or physics at school, almost everyone likely to read this book will have some recollection of (for instance) the meanings of such terms as 'angle' 'radius', 'numerator' and 'factor', and it would hence be wearisome to define every such word whose meaning is not clear from text or diagram.

Similarly, only typical examples are given of the elementary rules of algebraic manipulation. But unless the reader is quite happy with these it would be a good idea for him to work out a few examples of his own which will test such statements as that $a(b+c) = ab+ac$ by putting actual numbers in place of the symbols.

The handling of mathematical proofs raises a slightly different problem. A few have been included here (if sometimes not in their most rigorous of forms) for the benefit of those wanting to gather something of the 'feel' of the subject as well as its history. In that they represent a certain period or subject-matter, these demonstrations are typical; but in another sense they are not typical at all, for they must of necessity be unusually straightforward and self-contained examples. In some branches of mathematics the depth and value of a proof fortunately bears little relation to its complexity, so that in the theory of numbers (for instance) important theorems can be found which start from first principles and occupy only a few lines of a reasoning as much verbal as symbolic. But mathematical proof is not often as simple as the examples in this book may suggest, and there are whole branches of the subject where any worth-while proof would demand almost a chapter of explanation which itself expanded several pages of symbols and manipulations. In such cases the results stated must be taken largely on trust.

Discussions of applied mathematics also raise special difficulties, for to appreciate the greatness of a number of mathematicians involves some understanding of the advances made in various branches of physics by their contemporaries or by the mathematicians themselves. Many famous names from Archimedes to Einstein and beyond, indeed, belong to the history of physics as much as to that of mathematics; and it would be biographically absurd to mention Newton and ignore gravitation, or to speak of Gauss and not electrical theory, even were the physics and the mathematics of such men not in fact closely linked.

On the other hand, this book cannot hope to mention even every major *field* of physics and astronomy which has a substantial mathematical content; and so again only examples can be given of the numerous (and sometimes very unexpected) ways in which the world of pure logic and abstract symbols in which the mathematician lives interacts with the world of *things* which is the traditional concern of science. Some of these examples even come from the trivialities of ordinary life – but this is not to imply that the mathematics involved is itself trivial.

Almost every paragraph in this introduction could be expanded to chapter length; but though this book would be meaningless without some discussion of the ideas and philosophy of mathematics, its concern *is* with the mathematicians themselves – with the line of men who link the centuries before the alphabet was invented to our present age of computers and journeys into space. To keep a sense of perspective, it should again be stressed that compressions have had to be made here too; for it has been calculated that a reasonably complete (but still abbreviated) history of mathematics would occupy a library of over a hundred volumes, each longer than the present one, whilst several 'short' histories ending around 1850 are four times the length of this book. Similarly, nearly 10,000 men (and a few women too) are thought to have made lasting contributions to the subject, with even a 'short list' numbering over 500.

In these pages we can mention under a hundred pioneers, give real notice to only a dozen or so, and grant to less even than that number the type of attention (including contemporary quotations and human anecdotage) which brings a man to life. But there *are* a handful of mathematicians whose lives summarise a particular age and way of looking at the world, and these will be found dominating certain chapters here.

Finally, it may be that through its very brevity this book will present the history of mathematics more vividly than can the detailed and scholarly works on which it is so gratefully based.

I

Dawn in the East

In the beginning – wrote St John – was the Word; and whatever this means in terms of theology, one of mankind's greatest inventions was certainly that of language. Just as important, though – if coming much later in time – was the invention or discovery of *number.*

We do not know when this came about, but from the study of primitive tribes today we can guess that the first numbers were simply 'one' and 'many': it was probably thousands of years more before numbers greater than unity began to be given individual names, and thousands more again before a system was evolved which enabled men to describe *any* number, however large. What we can be fairly sure of, though, is that all these advances took place in the same part of the world. This was Sumeria, Babylonia or Mesopotamia, the basin of the Tigris and Euphrates rivers in what is now Iraq.

Here and in near-by Egypt was the heart-land of all the world's civilisation, the areas where Stone Age man first gave up the nomadic life of wandering with his flocks and settled down in villages and cultivated fields; and from here, dated some 4,000 years before Christ, come the oldest of written records. These documents, though, show men with so comparatively-advanced a mathematical system (it is far ahead of that of a typical African tribe today) that it is clear that a great break-through had taken place much earlier. And it is worth spending a moment on this break-through which may date back as far as 100,000 years BC, not only because it represents in itself one of the huge achievements of the human mind but because it has recently been seen to hold an important clue to the nature of number. It also illustrates that interplay between practical needs and deep ideas which is characteristic of the history of mathematics.

The crux of the matter is that man's first use for a number system was almost certainly to check on the animals in the flocks which he had learned to domesticate. Now, not only the behaviour of primitive tribes today but references in early literature make it probable that he often did this by keeping a store of as many stones as he owned beasts, and moving these from one pile to another as he gathered his herd in for the night. The device, of course, is still used by cricket umpires who have (usually) to count only up to six: a similar trick, which the reader probably used as a young child, is to use the fingers as counting-pieces.

All this may seem rather trivial; but it is of more than the historical interest marked by the fact that we still use the terms 'digit' (meaning a finger) and 'calculate' (from a word meaning a stone). For it establishes the idea of a *correlation*, a pairing-off of one set of things against another. There was, it seemed, something in common between the group of fingers on a hand and one possible group of a man's cows which could be expressed as 'fiveness', something which united a particular batch of apples and a particular cluster of stones which was comprised in the idea 'nineteen'.

Whether these counting-numbers or *positive integers* already in some sense existed in nature and only awaited man's discovery of them, or whether they were a creation of the human mind, is a question which has divided thinkers for thousands of years; and just as debatable, perhaps, is the relation of the integers to the idea of order or sequence and with this of time itself. But what is certain is that when men began to think of 'nine' as something meaningful without reference to a particular herd or heap, just as 'black' or 'heavy' was meaningful, then they had discovered an idea which has ruled mathematics ever since. Immensely more important than any particular theorem, this idea was that of abstraction and generalisation – of the tearing-out from a special situation of those elements in it which could be applied to any similar (or, sometimes, apparently *dis*similar) situation.

Another major Babylonian advance was the invention of a sensible system for naming – and writing-down – their numbers. This mathematical 'notation' began when men grew tired of counting with stones or on their fingers and realised that the same job could be done by making tally-marks in the dust or on a stick. Our present-day numerals are descended from the Babylonian ones, so that we can still recognise how the digits from 1 to 5 have been derived from such scratch-patterns. Above that the number of marks became hard to recognise, and more arbitrary symbols took over.

Obviously, though, men could only remember the shapes and names of a limited number of such devices, whereas (as the Babylonians realised) there was no imaginable limit to numbers themselves. The way out of this difficulty was to use a fixed number of symbols, and when one had reached the highest of them but still wanted to go on counting to make a modified mark (or give a modified name) signifying that a group had been completed and that one was adding to it by beginning at 'one' again. A completed group *of* groups would need yet another kind of mark – and so on.

The problem of notation was not to be fully solved until nearly AD 1000, and so we shall have to return to it. But, meanwhile, how large was the group of symbols to be? A sensible and useful counting system can be built up from any base or 'radix', and so the choice was simply one of accident and convenience. Too small a radix meant that largish numbers had to be represented by very long chains of digits: too big a one meant that a whole host of different symbols had to be invented, named and memorised.

In most civilisations the radix chosen was either ten (probably hit on from the habit of counting on the fingers) or twelve. The latter had the advantage that it could be roundly divided by three and was in fact favoured by the Babylonians – who also used the larger 'twelve-shaped' groups of 60 and 360 which, thanks to them, we employ today in our measurements of time and of angles. Similarly, we retain special words for 'eleven' and 'twelve' (rather than saying 'oneteen' and 'twoteen') and speak of dozens and grosses, whilst until very recently the 12 radix was reflected in Britain's coinage as well as weights and measures. For the pre-metric currency and metrological systems of all the world can be traced back to the work of the Babylonians.

But it was the 'denary' radix of 10 which eventually came to underlie the *counting* systems of civilisation, so that when (for instance) we today write 123 we mean three units plus two groups of ten units plus one group of ten times ten units. On a 12-fold or duodecimal system, incidentally, the same symbols would imply 3, plus 2×12, plus 1×144, or 171: on a 5-based system they would signify 3, plus 2×5, plus 1×25, or only 38.

These Babylonian mathematicians who have left us their records written on clay 4,000 or more years ago also realised that in some cases a meaning could be given to *negative* numbers: if '2' represented a man's wealth, for instance, 'minus 2' could stand for a debt of the same amount. Even negative lengths and times might have meanings as results to problems which had been worked out by rules which had been found to give reliable answers in other cases – the length, for example, being measured *backwards* from some implied starting-point, or the time being dated *before* a given event. But negative weights and areas seemed meaningless; and so the Babylonians – like the thinkers of several great civilisations after them – took 'minus' numbers less than half seriously.

Another mysterious-seeming kind of number also cropped up to puzzle these early mathematicians. They had discovered that quantities could be usefully and systematically combined in two ways – by *addition*, which was equivalent to taking two heaps of stones and simply mixing them together, and by the more abstract *multiplication* which represented the taking of a certain group of stones a certain number of times over. Two points are worth commenting on here – the recurring importance of the idea of *groups* in mathematics, and the fact that, while it seems obvious that three plus two is the same as two plus three, it is by no means obvious that two groups of three represent the same number of objects as three groups of two.

We shall have to return to such questions later. Meanwhile, though, it was becoming clear that the 'inverse operations' to adding and multiplying – i.e. subtracting and dividing – each held its mystery. The first of these was that, whereas 2 could be subtracted from 3 without any trouble, 3 subtracted from 2 gave a negative answer which might or might not seem 'sensible' depending on the problem which the numbers represented. And the same kind of difficulty appeared when a farmer wanted to *divide* his flock between two sons: if he had (say) eight sheep there was no problem, but if he had nine then one son – or one sheep – would be unlucky.

However, though you cannot have half a live sheep you *can* have half an acre of land; and as man increasingly became a farmer he had to pay as much attention to the endlessly-divisible world of lengths and areas (and the concept of area as a product of extension in two directions is another example of mathematical abstraction) as to that 'discrete' or countable domain of beasts and stones to which the whole-number integers were appropriate. In fact the idea of fractions or 'broken' numbers (which was also closely linked to the idea of ratio or proportion) turned out to present little real difficulty, and the Babylonians soon learned rules by which these could be added, multiplied, or simplified by – for instance – dividing their two parts by the same number or by removing 1 from those greater than unity so as to leave a 'proper' fraction.

Furthermore, it was obvious that a line could be halved: presumably, then, one could divide it into half-halves, and half-half-halves, and so on for ever. Alternatively – as when choosing a radix for counting *upwards* – one could decide to chop it into thirds, or tenths, or any other divisions one liked. This discovery in turn gave the Babylonians the idea that, just as any number greater than one could be expressed by so-many units, plus so-many groups of units, plus so-many groups of groups, and so on, so any fraction could be expressed as so-many divisions, plus so-many sub-divisions, etc.

For instance, to represent one divided by eight in our present system we can write the result as $\frac{1}{10} + \frac{2}{100} + \frac{5}{1000}$, or the decimal ·125. A full decimal notation did not come in much before the seventeenth century AD; and, as we have seen, the Babylonians in any case preferred the base of 60 (which itself survived until 1600) to our present 10. They also rather disliked fractions with numbers other than 1 in the top or 'numerator' position. But they moved towards, even if they did not completely grasp, the principle that any number could be expressed in the 'serial' shape of – for instance – $5 \times 60 \times 60$ *plus* 3×60 *plus* 9 *plus* $\frac{1}{60}$ *plus* $\frac{7}{60 \times 60} \ldots$

With all this, the Babylonians had an adequate 'kit of parts' for attacking all the problems in their lives which could be solved by mathematics. In fact, they began to go beyond such practical problems and to start exploring numbers for their own sake, either as an intellectual exercise or in search of mystical secrets. One direction this exploration took was the investigation of the special properties of positive integers – the study which a mathematician would today recognise as 'pure' arithmetic. Another was towards generalisation – towards the discovery of relations which held true *whatever* the numbers used. This kind of mathematics is today called algebra, and just how far the Babylonians advanced with it is a question which still intrigues scholars. But it is probable that much of the work which we shall be describing when we come on to the Arabs had its birthplace in that historic Garden of Eden beside the Euphrates.

The problems of measuring and dividing fields also led the Babylonians towards yet another kind of mathematics, that which dealt with relations in space. From solving simple but important practical problems such as the sharing of land or the levelling of a canal, this too developed into a mental exercise. But from its origins as an aid to surveying it is still called *geometry*, or earth-measurement.

Finally, the Babylonians combined their calculating skills with their knowledge of the ways in which (for instance) a flat or 'plane' surface could cut a sphere in another great achievement, the building-up of a rational science of the skies and the seasons. This study, indeed, was to dominate practical mathematics until about AD 1500. But astronomy itself is outside the scope of this book, and instead of being tempted into a discussion of the wonderfully accurate early work on establishing a calendar we should perhaps pause for a moment to look at the Babylonian mathematical achievement as a whole.

These men were shepherds, newly emerged from near-savagery. They had a religion, and in many ways an elaborate one; but it was based on blood-sacrifices rather than ethics.

Their arts too were complex rather than expressive. They could weave and they had invented the potter's wheel and the baker's oven, glass and perhaps a form of paper; but they were only beginning to evolve from the Stone into the Bronze Age, and their knowledge of metals was slight. Their technology was limited, their tools crude, their medicine mostly magic, their science almost non-existent. Above all they are, to us, *faceless.*

Yet such men (and in reality they were not nameless 'folk' or even lucky-chancers: there must have been geniuses among them) saw at least the dim outlines of almost every idea which lies at the heart of mathematics today. These concepts include the difference between the discrete and the continuous: numbers, operations and inverse operations: groups and series: the negative and the fractional: the treatment of integers, generalised numbers and space as different but closely-related studies: and above all the pressing of intellectual inquiry beyond the needs of day-to-day calculation. Most of these ideas, too, belong to the early period around 3000 BC, the epoch of Noah's flood. After this only a handful of major concepts had to emerge before mathematics could begin to take up the shape we know today – though for some of those concepts the wait was to be a long one.

After 1500 BC, Babylonian civilisation itself fell into decline. But mathematics was not forgotten, even if it was to make little advance for nearly a thousand years. For the bordering culture of Egypt had need of numbers too.

That civilisation was as old as the Babylonian (and may even have been older), and had its own achievements. But its people were on the whole less imaginative than their neighbours, and only the demands of a religion which was itself closely linked with astronomy, with the calendar, and hence with mathematics, took them far outside their everyday concerns. Foremost among those concerns, however, was the need to cultivate their crops. The whole culture of Egypt was based on the behaviour of a single great river, the Nile, and in particular on its yearly floods. These re-fertilised the land – but they also washed away most attempts at fences or boundary-marks, so that every peasant and tax-gatherer had to be able to establish his rights with a plan or a description of distances and angles.

As a result, a knowledge of at least rough-and-ready calculating spread from the priestly advisers of the Pharoahs and the engineers responsible for great pyramid and canal undertakings down to more ordinary people. Soon after 2000 BC, for instance, there could be found in almost every Egyptian village a 'rope stretcher' armed with a sighting-rod and measuring line who knew how to set out a right angle or multiply-up the dimensions of a figure to find its area – and who could probably also calculate the volume of solids such as a cone and use some simple algebra. The one field in which the Egyptians moved ahead of the Babylonians was such *mensuration*: whether by logic or trial-and-error, for instance, they found a close approximation to that very important 'constant', the ratio of the circumference of a circle to its diameter. And, as a memorial more important than any pyramid, the Egyptians left to later ages the world's earliest mathematical text – the 'Rhind papyrus', a document copied from older sources by a priest named Ahmes about 1700 BC, the time when Moses was leading his people out of captivity.

The twin cultures of Egypt and Babylon led the world in mathematics – as in many things – until about 1000 BC, when a new and more brilliant force began to be felt. But they did not exist quite in isolation. By 3000 BC, for instance, a civilisation had grown up in western India whose mathematics (at least) was so similar to that of the Babylonian school as to suggest that it was a direct descendant from it. In China, too, mathematics thrived before 2000 BC – though this knowledge, like that of the Maya culture of central America whose calendar-linked calculations may have a history as old as Babylon, was possibly of independent birth.

Perhaps we should note here that this book cannot pay fair attention to the achievements of oriental mathematicians, particularly those of the Hindu school in the first thousand years of the Christian era. During this period Indian mathematics was to become – except in one department – as advanced as its Mediterranean counterpart and often ahead in particular fields and discoveries, though these eventually filtered back to Europe. The story there is better researched and documented; but our main reason for concentrating on Western mathematics is that it was this which was destined to make a great take-off and to absorb all the world's mathematical thinking in the few centuries before the present international age. Today, of course, an advance may come jointly from a Japanese professor in San Francisco and a Cambridge man teaching in Bombay.

Finally, it is not clear how far westward the mathematics of the Tigris and the Nile penetrated. Certainly it spread along the coast of North Africa: to say more without doubt we would have to be wiser than we are today about the early migrations of mankind and the meaning of such structures as Stonehenge. But there are intriguing common features in the dimensions as well as the architecture of a whole series of stone monuments, built around 2500 BC, which extends from Ceylon to Ireland. And so it is not just fanciful to wonder whether, as long as 5,000 years ago, there were not priests in the Orkneys whose religion, astronomy, calendar and hence mathematics were directly linked to those of the Babylonians.

Another mysterious civilisation is that of Crete. Yet this island, along with the cities of Asia Minor, became a storehouse of mathematical knowledge when Egypt itself began to decline and civilisation awaited a great leap northward to the isles of Greece.

2

Archimedes and his World

At the start of *Men and Discoveries in Electricity* readers met one of the 'seven sages of Greece', a shrewd and successful business-man-turned-philosopher named Thales from Miletus in Asia Minor. Thales – who discovered that amber was in some way temporarily changed by friction, and who is important in the history of astronomy too – lived during a great era of the human intellect, about six centuries before Christ and at a time when the older civilisations had finally handed over their mathematical knowledge to that Greek one which had itself been created out of the energy brought down by invaders from the north; and he is thought to have discovered half a dozen geometrical 'theorems' (or proved results) which are taught in classrooms today. A typical example of these states that any angle which can be drawn in a semi-circle – say $\hat{A}$ or

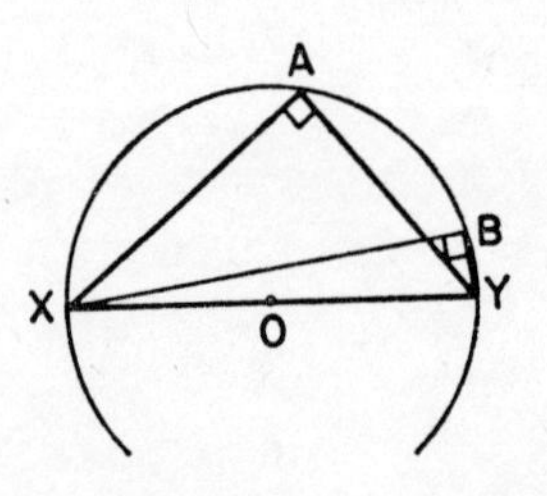

$\hat{B}$ in the diagram, where XOY is a diameter – is a 'right' angle of 90°. From our present viewpoint, though, the result itself is of less interest than the context in which it was deduced.

In the first place, it did not rise from nothing: anyone might

guess at this truth after some measurements, but to *prove* it demanded that several facts about triangles and circles should have been established earlier. And, secondly, it meant that men were now thinking about the properties of lines and points in an abstract way, concentrating on them as patterns rather than as a means of working out their tax liabilities. The conclusion *might* come in handy in surveying land or perhaps lead on to a more general result, but it was obviously discovered for its own sake and as part of a satisfying scheme.

This applies to almost the whole of the mathematics of the Greeks. Even in its purest form that civilisation endured for over 400 years, and perhaps the most remarkable thing about it is how little its ideas and ideals seem to have changed over a longer period than divides us today from Shakespeare's age. There were, of course, periods of swifter and slower progress; but from the moment when this culture – as different from any which the world had known before as it was from any which was to follow – first arose like a goddess from the sea, down to its decline about a century before Christ, the changes in thought and feeling are so comparatively minor that we can here jump across centuries and discuss themes without worrying about their strict sequence.

Almost equally remarkable is the *homogeneity* of Greek culture. This word, like so many used in the discussion of ideas, is itself a Greek one meaning 'all of a kind', and in fact it is impossible to discuss Greek mathematics without saying a certain amount about Greek life – just as it is impossible to say much about Greek life without introducing Greek mathematics. This chapter hence not only can but *must* leap to and fro in time, and it is not absurd to start a description of the 'Hellenic' period with the mention of a man who lived closer to its end than its beginning, Euclid.

For over 2,000 years after his lifetime around 300 BC the

name of Euclid was synonymous with geometry. And yet there is no certainty that he ever worked out a single theorem for himself. His achievement was the greater one of gathering together all the discoveries of a certain kind which had been made throughout a dozen generations, arranging them in an orderly pattern, filling in the gaps and – above all – reducing to a minimum and then identifying that part of the system which depended on intuition and could probably never be proved.

All the rest, he showed, could be built up step by step from a handful of *definitions* (e.g. of points, lines and circles), of *postulates* as to what could be done with these (such as that a finite straight line could be extended indefinitely), and of *axioms*. These last were appeals to common sense; and although a round dozen were needed, most of them were not really geometrical or even mathematical in nature. They stated, for instance, such 'self-evident' facts as that when equal things were added to (different) equal things, then the results were equal.

Euclid hence proceeded in the same way as Plato, Socrates and Aristotle had proceeded a little earlier, and asked only for a minimum of agreement about 'the laws of thought' before, in thirteen famous 'books' which were in fact parchment scrolls, he went ahead to build up a huge structure of ideas. The work of these philosophers themselves had been largely misguided because they did not realise the importance of experience as well as of intellect in constructing a comprehensive view of the world; but Euclid, operating in the limited field of one kind of geometry, was so successful that his books were used as standard school texts up to the present century and remain the classic example of the way in which to create '*a* mathematics'. In their time they have bored millions of young and unwilling readers; but against these we can count the thousands whose eyes were first opened to a new world by the elegance of Euclidean logic, rather as the poet Keats's eyes were opened when he looked into a translation of Homer. And some of these students went on to become the world's great mathematicians.

So we should here give at least one example of Euclidean geometry. If the reader is too near to the school-room and its (perhaps unimaginative) teaching not to shudder at the diagrams on the next pages, then he can skim over them without too great a loss. But there may be other readers ready to look at this kind of thing with fresh eyes, if only because an examination result no longer depends upon it. And some of these may themselves appreciate why Bertrand Russell found his first experience of Euclid as shattering as first love.

The diagram here shows two parallel lines (and though the idea of lines which always keep the same distance apart *seems* natural enough, it in fact holds the key to a great deal of geometry), with a third line crossing them. It may appear

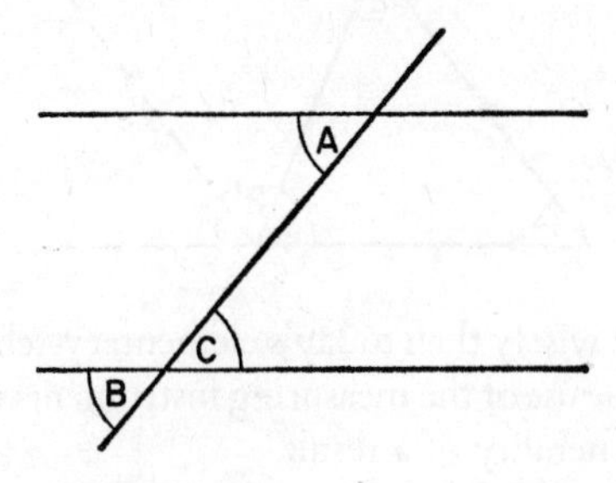

obvious – at least if the surface is flat – that the angles or degrees of turning marked $\hat{A}$ and $\hat{B}$ (which are called 'corresponding') are equal, and also that $\hat{A}$ equals the 'alternate' angle $\hat{C}$: or, if it is *not* obvious, we can surely check with a carefully drawn diagram and a scale of angles. But it is of the whole nature of mathematical proof that we are not allowed to use intuition and not allowed to make deductions from a few – or even a million – experiments. It is the great glory of Euclid and his predecessors that they realised this more clearly than did the Egyptians or even the Babylonians, who had resorted to measurement when logic failed them. The Greek dislike of experiment, which held back their science, was an incentive to their mathematics.

So Euclid saw that he must *prove* the equality of such pairs of angles. The proof demanded several intermediate stages, some of these being more complicated than one would expect and appearing useless in themselves. Indeed, even these 'pairs of angles' results look trivial – until we see what can be done with them.

The next figure shows a triangle – any triangle. What is the sum of the internal angles *1*, *2* and *3*, asks Euclid? And

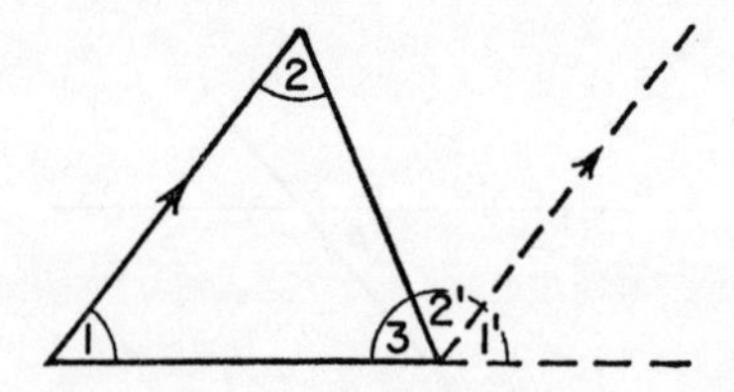

(perhaps more wisely than today's elementary-school teachers) he forbids us the use of the measuring instruments which would destroy the generality of a result.

Instead he extends one side of the triangle (in this case the base) as is shown by the horizontal dotted line – for in mathematical proof it is often necessary to introduce intermediate ideas before the final simplification. He then draws another new line (the slanting dotted one) so as to be parallel to the original side marked with an arrowhead: both these 'constructions' are specially allowed for by his postulates. Now, says Euclid, we have already shown that pairs of angles such as $\hat{1}$ and $\hat{1}'$ are equal, and also pairs like $\hat{2}$ and $\hat{2}'$. If we add to both groups the angle $\hat{3}$ we have $\hat{1}+\hat{2}+\hat{3} = \hat{1}'+\hat{2}'+\hat{3}$. But $\hat{1}'+\hat{2}'+\hat{3}$ make up the angle surrounding a point on one side of a straight line, which is itself half the total, 'round the clock' angle defined as 360°. And so the three internal angles of a triangle must also add up to 180° or two right angles.

What we have done here is prove a result which must be true for any triangle drawn on a flat surface – and we could never have achieved this degree of certainty had we spent a lifetime measuring-up *particular* triangles. The result we have reached (which is another of those which Thales may have discovered) is in fact an enormously useful one: all surveying by 'triangulation', or the splitting-up of areas of land into triangles, depends on it, and it leads on to results with even wider applications. But Euclid himself cared little for that and everything for the beauty of his proof, for he once told a pupil who queried the use of mathematics to collect twopence from a slave if he needed to be paid for learning. Geometry (as the competent mathematician Plato had pointed out) had come a long way from mere 'earth-measuring' now.

Of the men other than Plato himself whose work made possible Euclid's *Elements of Geometry* (which, in fact, contains a good deal of arithmetic), most are today no more than names. One Greek philosopher and geometer, though, stands out – the half-mythical Pythagoras, who lived near to the start of this period, studied – as did his tutor and neighbour, Thales – in Egypt, believed that number held the key to the universe, and may have been put to death for forming a secret society of mathematicians. Pythagoras is fittingly associated with what is perhaps the most important theorem in the whole of geometry – though one whose conclusion was known, at least in China, many centuries before Greece arose.

This theorem, which is illustrated a few pages farther on, states that in any triangle containing a right angle the length of the long side opposite that angle, multiplied by itself, equals the sum of the numbers obtained by multiplying the lengths of the two other sides by *them*selves. Rather irritatingly, no proof even today is both completely logical and elegantly simple; and we do not know how – or even *if* – Pythagoras himself arrived at a water-tight demonstration. But the theorem was so fundamental that two or three entire new branches of mathematics were later to be founded upon it.

Pythagoras' theorem can very easily be extended from the two dimensions of a plane into the three dimensions of ordinary space; and this introduces another aspect of the work of the Greeks, their discoveries in 'solid' geometry. To consider a typical problem in this subject, how many different kinds of solid are there which are 'regular' in the sense that a cube is – i.e. have all their faces, edges and angles equal? Pythagoras himself knew of five such, and thought this fact so important that it must be reflected in the musical scale, the movements of the planets and the nature of chemical elements.

This is an example of Greek 'science' at its weakest and most mystical, demanding that the real world should fit men's sense of what was appropriate. In contrast, the strength of Greek mathematics is shown by Euclid's proof that *only* five regular solids can exist, which he carried out by examining the possible ways of forming corners with equal angles. It looks simple enough now – but it was not so simple the first time round.

The name of Pythagoras provides a link to the other field of mathematics in which the Greek genius flourished. The ordinary day-to-day business of calculating prices, quantities of materials and so on was called by the Greeks 'logistics' (a word which today has assumed the quite different meaning of organising transport and supplies for large bodies of men): 'arithmetic' to them meant the investigation of the very nature of number, and no Greek philosopher ever confused these two studies as the English language confuses them today. Logistics, the Greeks believed, should be left to children, slaves and tradesmen: it was useful, but little more. But arithmetic – which today has to be called 'pure arithmetic', 'higher arithmetic' or 'the theory of numbers' – was real mathematics, something as exciting as a game and as beautiful as an art.

It is worth remembering here that it was the Greeks themselves who founded our modern concepts of both sport and art. The former had almost no links with mathematics other than that Greek geometry was very much concerned with obeying 'the rules of the game'; but the followers of Pythagoras tried hard to find numerical clues to the beauty and balance which they believed united the world. In poetry, for example, they drew up the laws of stress and rhythm which we remember today when we use such terms as 'Shakespeare's iambic pentameters', and in the visual arts they were delighted when they discovered that the proportions of length to breadth most pleasing to the eye (the 'golden rectangle', which is roughly the shape of a page of this book as well as of the Parthenon portico) seemed to be based on an interesting mathematical ratio.

In fact, even this claim is rather dubious and there are few other bridges between mathematics and visual beauty. In the third sector of the arts, however – music – the Greek approach was much more successful.

As we have seen elsewhere, the men who followed the

comparatively-practical Thales were held back from making much advance in the sciences by their prejudice against experiment. Though a gentleman was not allowed to measure speeds or to weigh metals, however, he *was* encouraged to play his flute or lyre. And so the Greeks naturally began looking for mathematical clues to the nature of music – and found them so easily and satisfyingly that it is surprising that they did not immediately abandon their dislike of making measurements. As it was, they failed even to realise that the fact that an organ pipe obeyed much the same rules as a violin string suggested that air was a real substance whose elasticity could be calculated.

We cannot go into this subject in any detail here: it belongs on the fringes of physics, far removed from pure mathematics. All we can say is that even before the age of Pythagoras it began to be realised that if one exactly doubled or trebled the tension on a plucked string – or shortened its length by a half or two-thirds – then one produced notes which blended well with the original one. The idea of *harmony*, which the Greeks felt must rule the universe, was for once well justified by measurements.

Today, almost all that is left of this kind of thinking is that the sequence of 'reciprocal' fractions: 1, $\frac{1}{2}$, $\frac{1}{3}$, $\frac{1}{4}$, . . . etc., is sometimes called a 'harmonic' series. But 500 years before Christ the type of number-mysticism which is nowadays the province of cranks (but which influenced highly intelligent men until the nineteenth century) was a cornerstone of thought. It is typical of the philosophy of Pythagoras, for instance, that he believed that the orbits of the heavenly bodies must lie in such ratios that they not only fitted in with the regular solids but emitted harmonious notes as they revolved.

Now we must look at the more substantial achievement of

the Greeks in general – and of the followers of Pythagoras in particular – in the world of number. This world was not a totally distinct one from that of geometry, though, for several arithmetical or algebraic truisms can be demonstrated by geometrical figures in which areas represent the products of multiplication. For instance, take two numbers, a and b, add

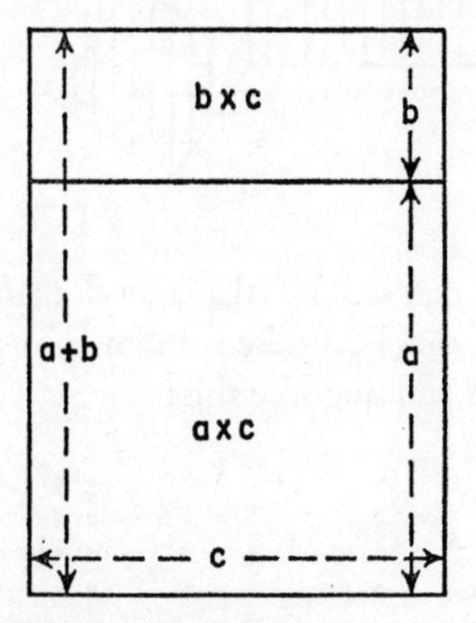

them together and multiply the result by a third number c. Is the answer the same as that got by multiplying a and b *separately* by c and then adding the products!

By comparing the areas in the diagram we can see that it is; and some rather more complex results were handled by the Greeks by the geometrical methods in which they were skilled rather than the symbol-manipulations which they found very clumsy. Conversely, we stated Pythagoras' own theorem a few pages back in the 'algebraic' terms of lengths multiplied by themselves. But its result is more in the form in which Pythagoras himself worked, which is that in the figure overleaf the two smaller shaded *areas* add up to the larger one.

A number multiplied by itself is thus still called a 'squared' number. (If we multiply it up again, we get a 'cube'.) This

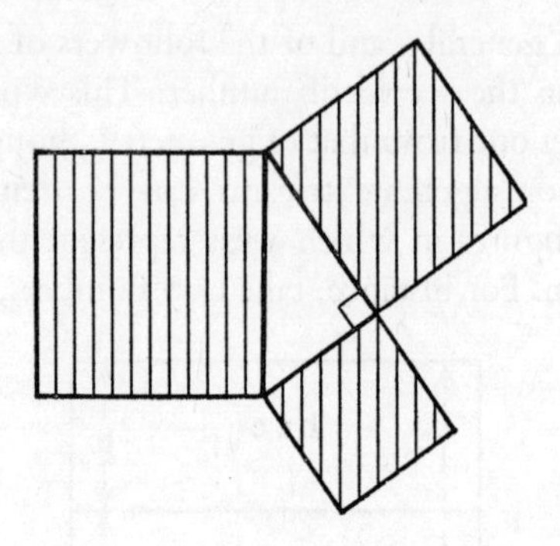

suggests a fact realised by the Babylonians – that certain whole numbers can be broken down into *patterns* of units, the 'square' ones running like this:

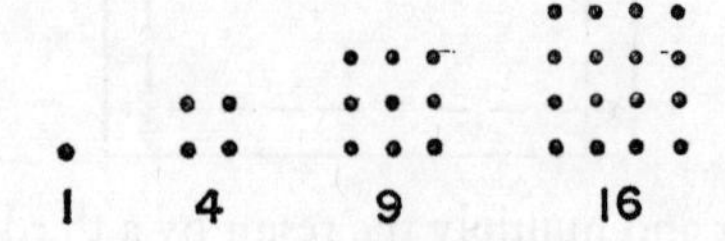

Now, how do we jump from (say) 4 to 9, from the upper figure below to the lower one? The larger number includes the smaller, whose side or 'root' is 2 and whose 'area' is hence 2×2. But to reach it we have had to add on the root itself twice over, once at the side and once at the top. And finally we have had to add an extra unit to complete the square.

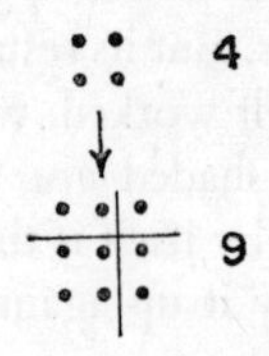

This in turn suggests that any 'square' number follows the pattern: $N \times N + 2N + 1$, where N represents the root of the square next below it. Put N equal to 4, for example, and the total comes to $16+8+1 = 25$; and 25 *is* the square of 5, the next integer upwards. But this test is a mere check on our reasoning, and that reasoning itself has done two rather splendid things. It has given us a formula by which we can build up an endless class of numbers of a special type, and it has reduced one kind of multipliction to a kind of addition. For the *differences* between successive squares run: 1, 3, 5, 7, ..., etc.

Continuing with this plan of 'geometrising' the integers into patterns of dots, another series studied by the Greeks was composed of 'triangular' numbers like those below. The reader might like (if he does not already know it) to work out a general formula for this series – i.e. for the sum of a chain of whole numbers such as $1 + 2 + 3$, etc: for it will appear later in this book. A clue is provided by the fact that – as the diagram shows – two successive triangular numbers add up to a square one.

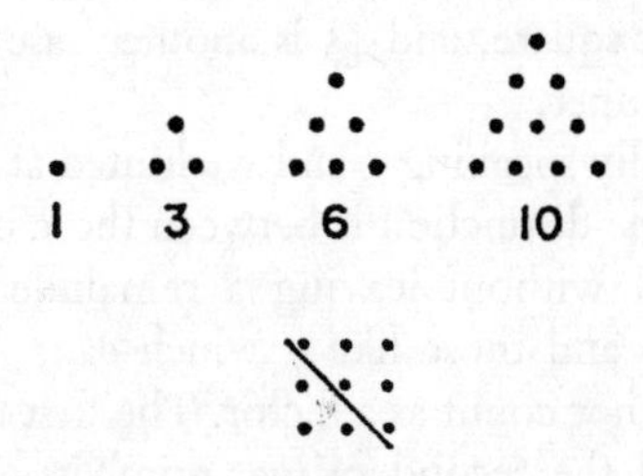

These are the kind of results which were achieved by the Pythagoreans, who were more interested in whole numbers than in numbers in general. Some of their proofs – though correct and neat in their way – were essentially makeshifts which could easily be bettered by algebraic methods which needed no reference to diagrams and which applied to fractions as well as integers. Much of pure mathematics, however, is concerned with the positive integers and also depends more on brilliant insight than on using an efficient 'sausage machine', so that other Greek proofs in this field remain as marvellous as do their counterparts in geometry.

Again, let us use the idea of presenting whole numbers as patterns. At the very simplest level we can recognise two types of integer – those like 4 and 6 which can be arranged as

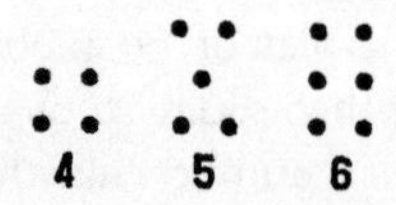

rectangles or 'oblongs', and those like 5 which cannot. Obviously any even number – except, perhaps, 2 itself – is 'rectangular', since it can be broken into at least two columns But (and one should never jump at 'obvious' converses in mathematics) this does not mean that all odd numbers are non-rectangular. For we have already seen that 9 is not merely rectangular but square, and 15 is another case of an odd but rectangular number.

Mathematically speaking – and we hinted at this idea in the last chapter – the distinction is between those numbers which can be divided without leaving a remainder into 'factors' (as $15 = 5 \times 3$) and those like 7 which cannot: *one* or unity of course, does not count as a factor. The first types are called composite, and the second prime, numbers. The difference between them is fundamental to the theory of arithmetic, for in the primes we have a series even more basic than the positive integers themselves. Indeed, just as we can construct any such integer by *adding* one to one to one and so on, so we can construct it by *multiplying* together a group of primes – unless it is itself prime.

Here (after 0, 1 and 2, all of which are special cases) are the first dozen primes: 3, 5, 7, 11, 13, 17, 19, 23, 29, 31, 37, 41. Obviously they slowly thin out as one progresses, since the higher numbers have more primes below them to act as possible factors. But even if we extend the series up into the thousands there seems on inspection little rhyme or reason to it.

The Greeks, however, believed in reason (if not rhyme) above all things, and were soon asking two questions. These were, firstly, 'Can we identify the primes, as we can (say) the triangular numbers, by seeing if they fit a simple formula or "pattern-gauge"?' And, secondly, 'If we cannot, can we at least say whether there is an infinite number of primes or not?'

The Greeks could not answer the first question – and we cannot answer it today. After 2,500 years in which almost every great pure mathematician has looked at the problem – and many have devoted years to it – we have progressed from a series of such tricks for 'sieving-out' non-primes as were invented by Euclid's pupil Eratosthenes (the great astronomer who was nicknamed 'Beta' or 'Number Two'), and can identify certain *types* of primes. It has also been a very fruitful search. But we still have no general solution.

The second question, though, was triumphantly solved by the Greeks even before Euclid's time. To repeat – we are asking a question about *infinity*, itself a concept which the Greeks usually preferred to leave alone. Yet the proof which Euclid quotes concerning the infinity of primes is not only unchallengeable but simple enough to be compressed into three lines.

It is perhaps not surprising that the Pythagoreans treated

their results in pure arithmetic as trade secrets to be concealed in jargon and never divulged to laymen; for a proof like this *does* resemble a conjuring trick. Once we know how it is done we lose (though falsely in the case of mathematics) a certain sense of wonder. But this book must not conceal anything; and so here is the lady sawn in half before the reader's very eyes.

Is there a greatest possible prime? asks Euclid; and in typical mathematical fashion he says '*If* there is, I shall call it *P*.' Then he asks us to consider another (and much larger) number made up by multiplying together all the primes from 1 to *P* inclusive and then adding 1 to the product. If we divide this new number by any prime up to *P* we shall, by its nature, get a remainder of 1. So either the big number is itself prime, or there is a prime between it and *P*. In either case we have contradicted our original postulate that there *is* a greatest possible prime; and so we must abandon this idea and accept that the primes are infinite in number.

Anyone who has a feeling for mathematics but has not met this theorem before may here feel like getting up to cheer: there was never a fighter's feint or a dummy pass at once so elegant and so effective. The method of proof used, though, is of a form in itself not uncommon.

There are no general rules for constructing mathematical theorems: this is part of the fascination of the art. However, there are one or two important *types* of proof, all of which were discovered by the Greeks. What we have just seen is an example of 'indirect' proof, or the demonstration of a fact by showing that the only alternative to it leads to nonsense. Another term for this is proof by *reductio ad absurdum*, reduction to an absurdity.

Some strict logicians dislike such proofs, saying that the fact that we have shown one course leading to a self-contradiction does not give us the right to assume that the only alternative which we can envisage is true: fortunately for them, indirect proofs can usually be reshaped into clumsier but more watertight forms. Most mathematicians, though, accept *reductio* proofs as being as valid as they are elegant. In effect, what the prover does in such a theorem is exactly what Socrates did in his moral speculations: he challenges the world to contradict him, *takes the contradiction at its face value*, and then shows that it contradicts itself and so leads to an absurdity. As one mathematician has pointed out, a *reductio* proof is like some magnificent chess sacrifice in which we apparently give away not just a piece but the entire game – in order to win it.

The primes themselves are important as being more 'specialised' numbers than the integers. By contrast, fractions are more 'generalised' numbers. Except for the 'reciprocal' $\frac{1}{2}$, $\frac{1}{3}$, $\frac{1}{4}$ ... type (which could be regarded as the inverses of integers) which were useful in the study of music, the Greeks were not very interested in them. But they recognised that *any* number defined by the ratio of two integers – whether it was as simple as $\frac{1}{2}$ or as elaborate as $\frac{121}{360}$ – was just as sensible, manageable and 'real' as an integer itself. It was simply that the Greeks preferred to deal with whole numbers, just as they preferred to think of a universe composed of atoms rather than of continuous fluids.

Early in their journey of mathematical exploration, however, the Greeks came across a quite different type of fraction. The matter arose like this.

We have seen how for every number there is a corresponding 'square'; 9 corresponds to 3, for instance, and 16 to 4. We have also introduced the inverse of this idea by using the term 'root'. The root of a number (or, more properly here, the *square* root – for the Greeks recognised cube roots too) is simply that number which, if multiplied by itself, would produce the other as 3 produces 9. Thus, 4 is the square root of 16.

Fractions as such do not lead to any special difficulties here, and our fallible commonsense tells us that every number must have a square root. But when we look at the sample even of whole numbers between 1 and 10 we find only three cases – 1, 4 and 9 – where we can name it straight off. Presumably 2, 3, 5 and so on also have their roots, though. As an example let us look more closely at the root of 2, for convenience using the modern notation (which in fact belongs 2,000 years later in time) of writing this as $\sqrt{2}$.

We know that this number must be more than 1 and less than 2. If we split the difference and multiply $1\frac{1}{2}$ by itself we get $2\frac{1}{4}$, so $\sqrt{2}$ is less than $1\frac{1}{2}$. A second trial will show that it is greater than $1\frac{2}{5}$. Changing over to today's decimal notation of expressing fractions as chains of tenths, hundredths and so on, we can go on narrowing the gap until we become tired after showing that the elusive number is a trifle over 1·1412... .

This is accuracy to a few parts in 100,000, as good as would be needed for almost all practical purposes. But mathematically we have not begun on the problem. In fact, we have not even shown that $\sqrt{2}$ 'really' exists.

However, this can be demonstrated in true Greek fashion by a geometrical 'model'. The diagram here shows a right-angled triangle with its two shorter sides each one unit long. Since one multiplied by itself equals one, it follows from the invaluable theorem of Pythagoras that the square of the length of the long side is $1+1$, and hence that this side itself measures $\sqrt{2}$ in whatever units we are employing. By drawing very accurate figures, we could thus confirm our arithmetical estimate: more important, we have shown that $\sqrt{2}$ is at least 'real' enough to be measured with a ruler.

Yet, whatever methods they used, the early Greeks were

unable to show that it was equal to the ratio of any two integers. Indeed, they came to have a strong suspicion that it was a completely different type of fraction. This proposition itself, however, was something which might be proved or disproved. And it was firmly proved in the age of Pythagoras

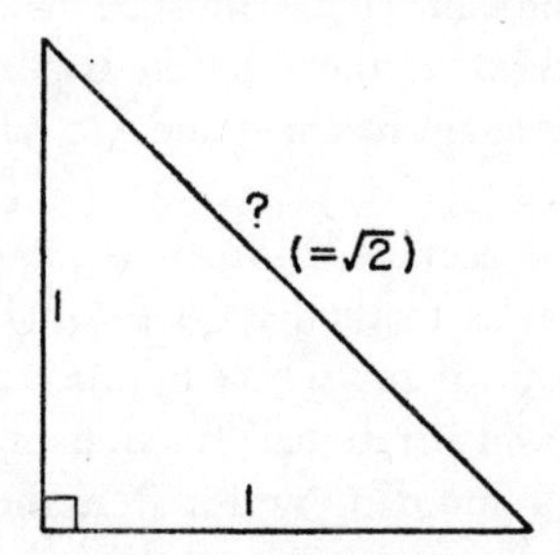

himself, of that great believer in the almost-divine quality of the positive integers.

The proposition we are faced with is that no ratio equals $\sqrt{2}$. To prove so negative a statement seems at first sight impossible; but the proof known to Aristotle and given by Euclid in the section of his *Elements* which deals with arithmetic is as elegant and unchallengeable as that of the infinity of primes. It is, though, a little more complicated, and can only be easily followed by employing methods which we have not yet introduced because they were not used by the Greeks themselves.

For those willing to cheat slightly by borrowing ideas from the next chapter, however, Euclid again challenges us, this time to produce a fraction such that $M/N = \sqrt{2}$: he demands only that we have reduced M/N to its 'lowest terms', so that if – for instance – we think $\frac{14}{10}$ is the answer we should present it as $\frac{7}{5}$. Another way of writing the above claim is to multiply each side of the equation by itself, so that

$M \times M / N \times N = 2$, or $M \times M = 2 \times N \times N$. Now, since the right-hand side has a factor of two and only even numbers can have even squares, M itself must equal twice some other number. Call this m: then, since $M = 2 \times m$, $M \times M = 4 \times m \times m$. Halve both sides of the new equation, and $2 \times m \times m = N \times N$, so by the same rule N must be even. Thus N and M both have factors of 2, the fraction was *not* in its lowest terms, absurdity is again reached, and $\sqrt{2}$ *cannot* be expressed in the form M/N.

This argument was extended from $\sqrt{2}$ to other square and cube roots by Plato's mathematical tutor. In fact, it can be shown that any root of any whole number, if it is not itself a whole number, is an 'irrational' fraction or one whose two parts of numerator and denomintor seem inexpressible in the same scale. But the discovery of the nature even of $\sqrt{2}$ horrified the Pythagoreans, for here was a quite simple number which would not bow to their sense of proportion. For instance, how could one multiply $\sqrt{2}$ by itself – though one knew the answer must be simply 2 – if one could never pin down the root in the first instance?

There were, admittedly, plenty of fractions which would not 'come out' roundly in a particular number system; thus $\frac{1}{9}$, expressed as a decimal, equals $\frac{1}{10}+\frac{1}{100}+\frac{1}{1000}\cdots$, or 0·111 ..., going on for ever. But the Greeks were clear about the difference between mere number-tricks (such as the rule that if – in our notation – the final digit of number was 0 or 5 then it was divisible by 5) which applied only to *decimal* arithmetic, and more fundamental properties. And it was clear that with $\sqrt{2}$ and the like they had to deal with a whole new type of number, one which in *any* system could only be expressed by some kind of infinite chain giving ever closer approximations.

It was not enough for the Greeks to be able to *handle*

numbers as brilliantly as we have seen them doing: they had to *understand* them too. And so, after having sacrificed a hundred oxen to propitiate their gods, they banished $\sqrt{2}$ and similar 'incommensurables' to the mental leper-colony where the negative numbers were already pining. The result still shows in our language, for we use the word 'rational' (which simply implies 'not accurately expressible in a finite number of digits') as a term of *moral* approval meaning 'reasonable'.

The real trouble (as we have already hinted) was that the Greeks were alarmed by infinity – and by its opposite, the infinitesimal or unthinkably small. To a degree they were right in this feeling, for infinity has to be treated – at least by mathematicians – with great respect. But it meant that as early as the fifth century BC thinkers became plagued by makers of riddles and paradoxes, such as the ingenious Zeno with his apparent proof that (for example) the fastest runner could never overtake a tortoise if the tortoise had been given a head start.

So far we have taken a glance at the two great fields of Greek mathematical achievement – pure geometry, which had comparatively little use for numbers, and pure arithmetic, which was concerned with the very nature of number though not with calculations. One thing these studies have in common is that both are in a sense 'useless' – the former largely, and the latter almost entirely, so. They have little to do with crops and calendars, ship-steering or temple-building. They stand or fall on their own, as does an ode or an Olympics; and their main influence is in generating more mathematics of the same type.

But we have also seen that there were serious weaknesses to Greek mathematics – its adherence to narrow rules, its reluctance to learn from the physical world, its inability to cope with infinity. These drawbacks had hindered all the philosopher-mathematicians who worked in the line which descended from Thales and Pythagoras: one of the few exceptions, perhaps, was the heretical Archytas, the founder of that study called mechanics which is intermediate between mathematics and physics. Archytas was also a 'mechanic' in another sense, for he loved making toys for children and seems to have been the inventor of something which Aristotle regarded as a great boon to the parents of active youngsters – the baby's rattle. He then turned to pleasanter sounds and became the first to realise how wind instruments worked.

Even before the time of Euclid, though, it had become clear that there was need for a man who combined the strictness of the traditional Greek outlook with a wider vision. What could not be anticipated was that this man should also be one of the world's overpowering geniuses. But such is the distinction of Archimedes. With his freshness of approach and his

willingness to tackle problems which did not yield to established methods, Archimedes would be remembered and revered simply as the first great name in the tradition which runs parallel to that of mathematics-as-pure-thought, the tradition of mathematics-in-the-service-of-science. But he was more than that: he was the possessor of one of the score, or dozen, or perhaps half-dozen most penetrating intellects that mankind has ever known.

Archimedes was born in 287 BC, in the Greek colony of Syracuse in Sicily. His father was a wealthy man and also a distinguished astronomer, and as a typical Greek philosopher Archimedes himself was to do important work in this science. For a while, though, he left Italy to study in the even more important colony of Alexandria. Here in the delta of the Nile Alexander the Great – the pupil of Aristotle, and one of the world's most civilised conquerers – had established a new dependency of Greece some fifty years earlier; and thanks to the wisdom of its rulers and their endowment of a great library, Alexandria had before 250 BC became an intellectual centre challenging Athens. In fact, Euclid himself had taught for some time here on the African coast.

Archimedes' own work was carried out under the patronage of the king of Syracuse, to whom he was probably related. It divides into two parts, the practical and the theoretical; and much of the first can be quickly disposed of if only because it is so legend-shrouded. Any man with a reputation for inventiveness living in an age when neither such men nor written records are common tends to be credited with all the bright ideas of his time, and Archimedes has hence suffered the same fate as Roger Bacon. It is improbable, for instance, that he invented the incendiary mixture called 'Greek fire', whilst two other 'secret weapons' attributed to him – a crane for lifting ships bodily out of the water, and a huge mirror for setting them on fire – sound in themselves rather far-fetched. All that these traditions prove is how early in history war acted as a spur to invention.

There is no reason, though, to doubt that a type of screw-conveyor still used for lifting powders was devised by Archimedes; and though this is not an invention of the first importance it shows that the Alexandrians were more mechanically-minded than the mainland Greeks had been. Vastly more important, both in itself and as an example of the analysis of the physical world, was Archimedes' realisation that every substance had its characteristic density (i.e. the ratio of the weight of a given volume of it to that of the same volume of water), and that when a body was completely immersed in water it displaced a volume of liquid equal to its own. This volume could be either measured directly or calculated from the difference between the weights of the body in air and in water, so giving the density.

The tradition is that Archimedes was led to this discovery by his king, who wanted to know if the court jewellers were cheating him by diluting his gold with the less dense silver: that the answer came to the philosopher as he stepped into an over-filled bath and saw how the volume of his body made it overflow: and that he then tore through the streets of Syracuse dressed only in a towel (if that), shouting 'Eureka', or 'I have solved it.' Some of the details may be over-picturesque. But there is no doubting Archimedes' genius for solving physical problems, even if the purely mathematical content of this one (as of his work on simple machines like levers) is slight and even if – in true Greek fashion, and perhaps with tongue in cheek – Archimedes claimed to despise such things and much prefer his work in pure mathematics.

To appreciate this work we must take a backward glance at a matter which had engaged Egyptian surveyors and Babylonian astronomers – the question of the ratio of the circumference of a circle to its diameter. If this problem stood on its own it would be a mere curiosity of mathematics – though still an important one, since the circle is the simplest of all curved figures. But in fact, as soon as men began to calculate not only the *lengths* of the boundary-lines of circles but such equally practical matters as the *areas* of the surfaces of cones and cylinders or the *volumes* of spheres, they found that the easier part of their work was in each case to derive a formula involving that crucial ratio. The ratio itself remained a challenge – and became so important that it eventually achieved the distinction of being identified by a special symbol, π.

This – the Greek letter 'p', short for 'periphery' and pronounced 'pie' – represents the length of the *curved* line bounding a circle as a multiple of a *straight* line through its centre. The Egyptians, at least, were not worried by the fact that there is an important logical difficulty concealed in this definition, and were mainly interested in π for practical purposes. Simple sketches of circles and triangles suggested that it was 'a bit more than 3'; but this estimate, though quoted in the Bible, is too rough for even the roughest of builders. The Egyptians themselves at first put π equal to $3\frac{1}{7}$ (note the typical preference for expressing fractions with a unit numerator), which is accurate to better than one part in 1,000. But in their pyramid-building they used straight-line standards good to at least one part in 10,000, and there is some evidence that they were inspired to try to *calculate* π too to a greater accuracy than could be attained by tracing large figures in the sand.

If this is so then they must have used a method which is historically attributed to Eudoxus, a pupil of Plato who worked his way up from extreme poverty to become an outstanding astronomer as well as a first-rank mathematician. Nearly 2,000 years after the time of Archimedes, the extension

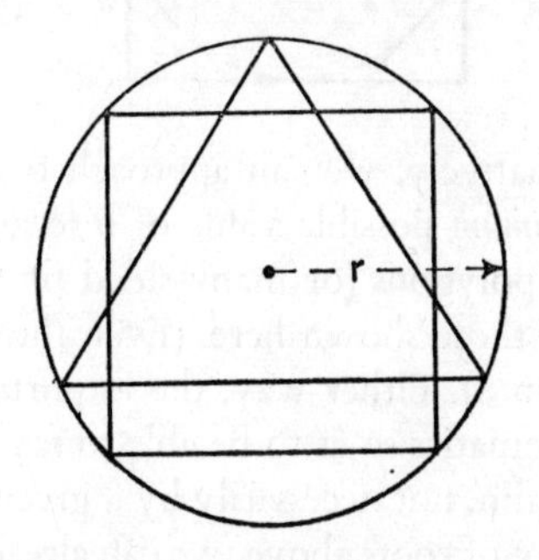

of this method of successive approximation which had also been applied to evaluating (for instance) $\sqrt{2}$ was to lead to a whole new type of mathematics; and so we can only introduce it now. But it is clear that the mathematician has to find a method of reducing a curve like the circle to a series of straight lines.

The illustration shows a circle with a triangle drawn inside it, and with little more than the theorem of Pythagoras we can show that the three sides of this triangle add up to exactly $3 \times \sqrt{3}$, or about 5·2, times the radius or half-diameter. This is a badly undersized 'fit' to the circumference, though, and we will get a better one by considering the 'inscribed' square. The same kind of Euclidean geometry shows that the sides of *this* add up to precisely $4 \times \sqrt{2}$, or about 5·7, times the radius.

Consideration of a six-sided figure shows that 2π is greater than 6, and from a twelve-sided one we can get a still closer

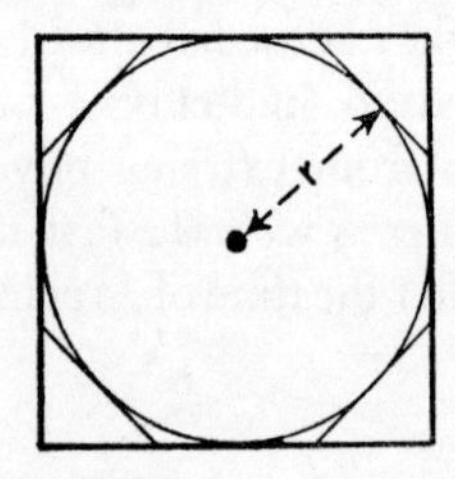

minimum. Alternatively, we can approach from above and see how the *maximum* possible value of π lessens as we consider the regular polygons (or many-sided figures) 'escribed' to a circle, as are those shown here. (Even the square proves that 2π is less than 8). Either way, the important thing is to arrange the mathematics so as to be able to say that "π is ..." and follow this claim, not necessarily by a given number (for, with all the display of roots above, we can already hazard that π is at least as irrational as $\sqrt{2}$), but by the suggestion of a *method* which will allow us to estimate the ratio as closely as we like.

For Eudoxus, Archimedes and the next in the main line of Greek mathematics, Apollonius of Perga, this method had to take a purely geometrical form; for instance, Archimedes himself improved on his predecessors by considering a ninety-six-sided polygon. But a few centuries later means were known which were close to the modern use of 'series', and to round-off the story we can perhaps look forward to the years after 1500 when more convenient devices were worked out.

For instance, an English eccentric in the seventeenth century achieved about 150 digits and had them inscribed on his tombstone – perhaps in tribute to Archimedes, who asked that *his* grave be crowned with a mathematical diagram. Finally, a few years ago a machine working for eight hours did what would have taken a man 30,000 years and calculated π to more than 100,000 'places'. For practical purposes, of course, nobody is interested in knowing much more than that $\pi = 3{\cdot}14159\ldots$. But these weighty calculations were not made just to show what could be done, but in the hopes that the expansion of π would throw light on some mysteries of number.

Even to Archimedes, though, π meant more than the ratio of the circumference of a circle to its diameter; and he went on to investigate its properties in the same spirit of confidence in which, following Eudoxus, he had partly resolved the philosophical difficulties posed by the irrationals and by the paradoxes of Zeno. We have already mentioned, for instance, the problem of finding the area enclosed by a circle. This was attacked by Archimedes and his colleagues in a similar way to that in which they had approached the cirumference problem – by splitting-up the whole into units, based on straight lines, whose total area approached that of the curved figures as their individual sizes shrunk and their number grew.

To ask the areas of *any* curved figure, like the one overleaf, is to ask a question which can only be answered approximately and by experiment – for instance, by drawing in on squared paper and counting the squares more than half covered. But we can reach a much more precise answer if we know that the figure obeys some law or rule which can be put in mathematical terms.

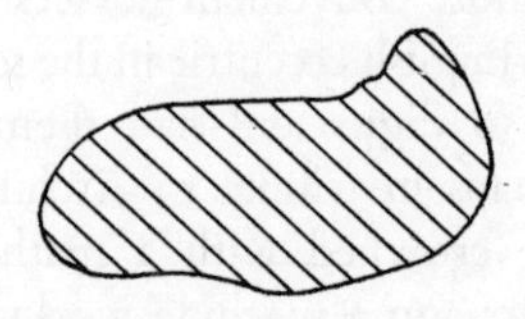

In the next diagram, two ways are suggested of splitting-up a circle into slices, the total area of which will be a little smaller

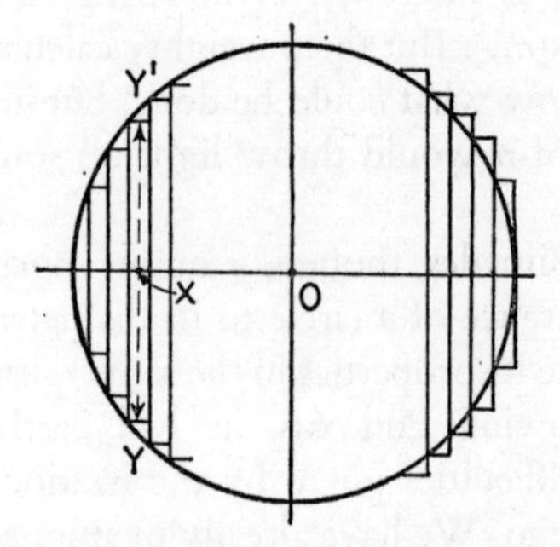

than that of the circle itself as drawn on the left and a little greater as drawn on the right. In either case the *limit* approached by taking an ever-larger number of ever-thinner slices should give us the area of the circle itself. The first essential for performing the calculation is hence a formula which relates the height of each thin rectangle (say, from Y to Y′) to its distance from the centre (O to X) – and this formula was given by the same 'spin off' from Euclidean geometry which enabled sailors to calculate how far out at sea the great lighthouse at Alexandria could be seen.

Archimedes and his school again carried out the considerable tasks of mathematical manipulation needed to express everything in terms of simply-measured quantities (like the radius of a circle) *plus* the universal constant π. But their real advance over the Pythagoreans was to appreciate that an *infinitely* long series (like that represented by the slices of a circle) could 'converge' to a *finite* limit, just as $1+\frac{1}{2}+\frac{1}{4}+\frac{1}{8}\ldots$, however many terms we take, will never add up to more than 2.

It was natural that the circle, that most 'perfect' of all figures, should especially fascinate the Greeks. This specialisation in fact went so far that Euclid was not only reluctant to analyse any other curve but had laid it down that none other should be constructed in proofs according to his (or Plato's) methods. However, there were several types of curve known to be at least as interesting as the circle.

There was, for instance, the *cycloid*: this was the curve or 'locus' (an idea due to Thales) traced out by a point on the rim of a wheel, and it had such relatives as that responsible for the

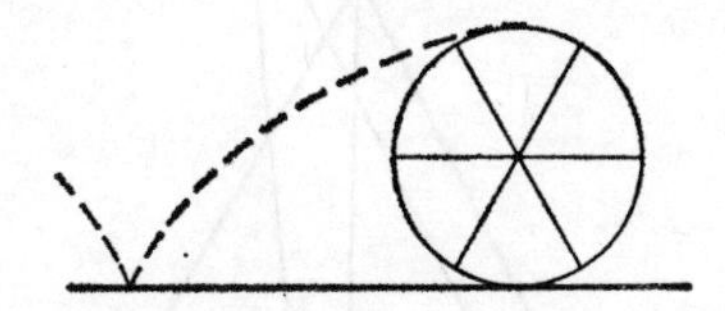

paradox that parts of an express train are always moving backwards. There was the whole family of spirals to which Archimedes devoted so much attention that the simplest, in which the turns remain the same distance apart, is still called an 'Archimedean' spiral. And so important that even Euclid had had to write about them were the curves which made up the group called 'conic sections.'

Archimedes had to leave his work on these to be completed

by Apollonius (who named them and wrote a fine book on the subject) and by Apollonius' own successors like Hipparchus. But before we look at the nature of these curves we should note that most of their investigation was possible only to men who had set aside the Pythagorean anxiety about dealing with infinity. For the infinitely-small was implicit in the nature of square roots, let alone of π.

Conic sections are just what their name suggests – but we must start with a mathematical cone, which is an ordinary one

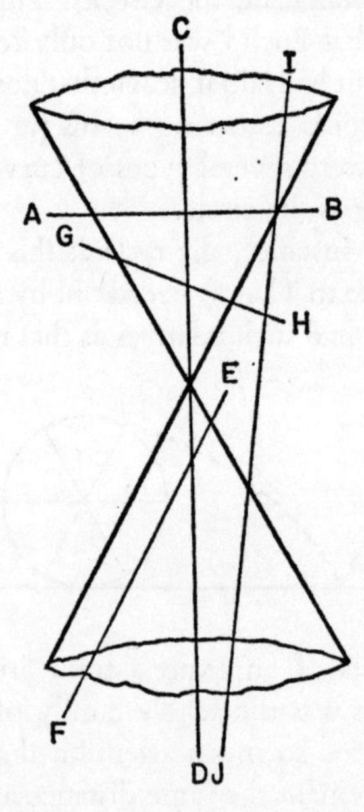

doubled and extending indefinitely far in both directions. This – as is shown here – can be sliced in five main ways. Cut straight across, as along AB, its section is a circle: in fact, in moving on from the circle to other conic sections Archimedes was following the historic stream of mathematics, which is to progress from the particular to the general. Cut the cones through their meeting-point, as along CD, and they yield another extreme case, a pair of straight lines. The third example is produced by slicing one along a plane off-centre but parallel to a side like EF. The result then (and it is shown on p. 53) is a *parabola*, a figure whose two arms curve more and more gently inwards and towards the state of being parallel to each other.

A gently-slanting cut like GH produces the more-or-less flattened circle mathematically known as an *ellipse*: and finally a sharper off-centre cut like IJ leads to the *hyperbola*. This curve is derived from both parts of the double-cone, though it is usually necessary to consider only one half of it. Its pairs of arms never approach the parallel condition, but they *do* approach a pair of crossed straight lines. The diagram here shows the most important case, that of the 'right' hyperbola.

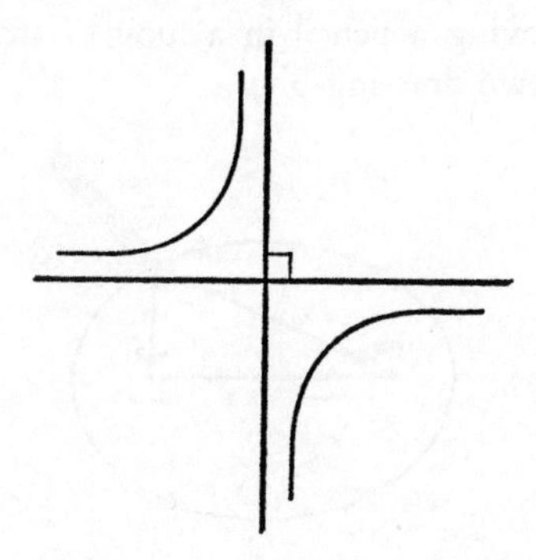

These three curves – parabola, hyperbola and ellipse – challenge the circle in practical importance. They can be seen everywhere in shadows and oblique views, but are even more often met with as *loci* (or the tracks of moving points, such as the cycloid above) than as visible shapes. Planets and other satellites, for instance, move through space in ellipses (and not in circles as the Greeks believed), and comets often follow hyperbolas. As for the parabola, we shall later find it describing the motion of all kinds of projectiles from cricket balls to rocket missiles.

The Greeks, however, were more interested in the purely mathematical properties of such sections: and these can be expressed in two forms, the algebraic one which we shall encounter later and the geometrical one which was used by Archimedes. This itself was based on the idea of a locus: for instance, a circle can be described as the locus of a point which moves so as to be always the same distance from another, fixed point – its centre. The ellipse is one step more complicated, for in this case the point which defines it moves so that the sum of its distances from *two* fixed points (themselves nowadays called the *foci*) is constant. This is why an ellipse can be drawn by moving a pencil in a loop of string which also passes round two drawing-pins.

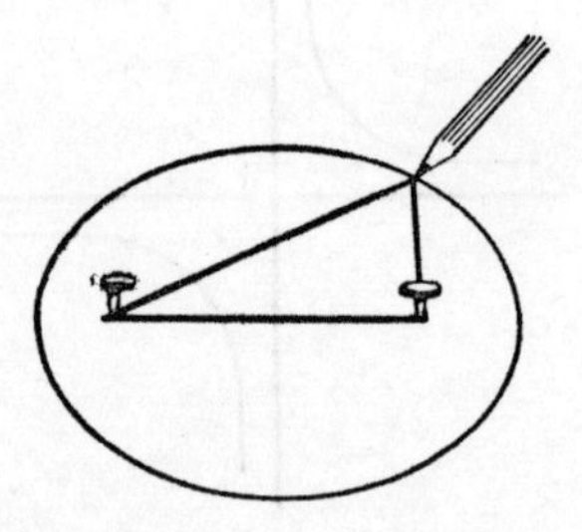

Another outcome of this property of the ellipse is that its two key points are foci in the optical sense too: all light-rays leaving one of them will meet again at the other. And a similar fact is responsible for the most useful property of the parabola.

The rule here is that the curve is the locus of a point which travels so that it is always the same distance from a fixed point (F – again called the *focus*) as from a straight line (AB): i.e. wherever P is on the parabola in this diagram, FP equals PD.

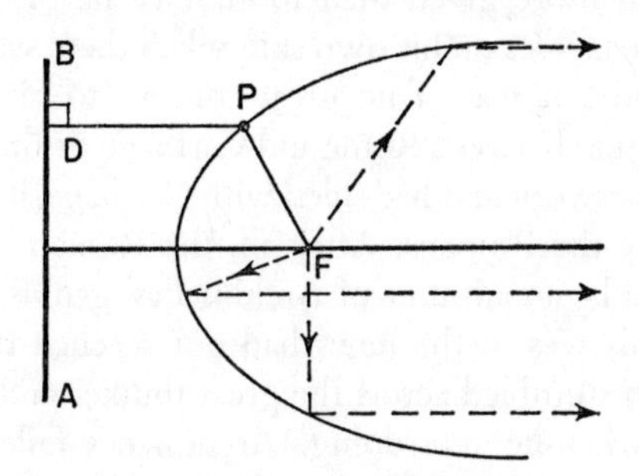

Now, it can be shown that any line leaving F – for instance, the three which are arrowed here – which 'bounces off' the curve at the same angle as it meets it at is parallel to all other such lines, including the horizontal axis. So, if we bend a mirror into the shape of a parabola, it will send out a parallel, 'focused' beam from a light source at F.

All this follows from the geometry of the parabola, so that even if Archimedes never made a parabolic mirror his work was awaiting when men needed searchlights and car headlamps. The reflectors of these are, in fact, parts of figures obtained by *rotating* parabolas about their axes; but Archimedes himself had gone on confidently from his work on plane curves to explore the corresponding 'solids of revolution' which were related to them as a sphere is to a circle. Among his most remarkable work, indeed, was the calculation of the 'centres of gravity' – the points where one could regard the weight as being concentrated – of such figures.

It was probably on such problems of statics (or the science of balanced forces) that Archimedes was working one warm day in 212 BC. He was sitting in the open and peacefully drawing figures in the dust – a common habit among mathematicians in the Mediterranean world, but one to which Archimedes was perhaps more given than most since he often sketched circles and triangles on his own skin when there was no better surface handy. It was of no great concern to him then that war was raging between Rome and Carthage or that Syracuse, which lay between and had sided with Carthage, had just been captured by the Romans. After all, the Roman general was supposed to be an admirer of Archimedes' genius.

But if this was so the news had not reached the illiterate soldier who stumbled across the great thinker and demanded to know what he was doing. Archimedes failed to show sufficient respect, the soldier drew his sword, and seconds later the body which had housed a mind not to be equalled in its field for nearly 2,000 years lay dead in the Sicilian sunlight. It was buried at the gate of the city; and there, more than a century later, the great Roman statesman Cicero found the tomb with its geometrical carving so neglected that he ordered it to be rebuilt. At the time of his death, Archimedes was seventy-five years old.

3

In Araby and Europe

In mathematics, as in science, the Romans added little to the Greek achievement. Their great engineering works – and their banking system – were extended across Europe without more than rough-and-ready calculations; and in all the 500 years during which Rome ruled our world she accomplished little more than a certain streamlining of the 'logistic' or figuring type of mathematics and some sound work on the calendar. It is significant that the Romans invented a goddess of mathematics, Numeraria – and then prayed to her mainly that she should make them rich.

But (as when the Egyptians followed the Babylonians) little was forgotten either. This is shown by the fact that the vocabulary of mathematics – and of geometry in particular – is today more Latin than Greek. Terms like 'quadrilateral' and 'Q.E.D.', in use alongside such Hellenisms as 'polygon', prove that the Romans studied their Euclid even if they did not add much to his work.

For more creative minds we must look across the Mediterranean; for long after Greece itself ceased to be an important power Alexandria remained intellectually alive. The potentialities of the geometrical approach of Euclid (and even of Archimedes) seemed to be becoming exhausted now, but the theory of numbers remained a challenge to the liveliest minds. And if the Alexandrian school produced no mathematical genius of the first rank for nearly 500 years after Apollonius, it is reasonably represented by Hero – an Egyptian who, about AD 50, devised the still-powerful formula by which the area of any triangle can be worked out from the lengths of its sides, who made a discovery of the first importance about the paths of light rays, and who is also famous for inventing a primitive steam engine. Then, perhaps two centuries later (for all that we know of his life is typically compressed into a kind of equation which says that he lived for eighty-four years), came the really distinguished Diophantus.

We have seen that Greek geometry itself had a numerical aspect. This had been somewhat neglected by Archimedes, but Diophantus realised that the theory of Pythagoras led to the question of what pairs of square numbers could be added together to form a sum which was itself a 'perfect' (or integral) square. The simplest case was 3, 4 and 5, since $9+16 = 25$. But we have to go up to 5, 12 and 13 – where both sides of the equation equal 169 – before we find another such example.

Useless as this kind of investigation appears, it is more than just a guessing-game and can lead into deep seas of thought. Diophantus and his fellow-Alexandrians not only knew of formulae for constructing an infinite number of such 'Pythagorean triads', but also considered problems which involved – for instance – three squares adding up to a fourth one. Studying questions of this type, they solved many and left more to baffle mathematicians up to the present day, when the solution of any equation in the form of integers is still called a 'Diophantine' problem.

One reason why the Alexandrians were so successful is that the solving of this kind of puzzle depended largely on brilliant 'one-off' ideas, and that until the end the Greek mind was fertile in producing these. To devise *general* methods for the solution of mathematical problems, by which even an uninspired thinker could produce valuable results, implied by contrast the invention of quite new mathematical instruments.

Diophantus himself had some idea of these implements. But

it was left to the next great civilisation to sharpen the tools needed in ordinary mathematical carpentry rather than these equivalents of a brain-surgeon's lancets which were useful only in highly-specialised operations. This later culture just overlapped with that which had ended at Alexandria after the death of the last great geometer of the ancient world, Pappus.

There, at the centre of a crescent which enclosed the eastern and southern shores of the Mediterranean, Greek concepts had met the residue of mathematical knowledge of the desert races – and, in so doing, had led to a clearer idea of the movements of astronomical bodies than men had had before. Another contribution had come from the Jewish people, especially after they had been displaced from their homeland by the Romans; and in the long centuries after Christ's crucifixion there was also commerce and conquest along the routes to the east, bringing mathematical ideas back to the Mediterranean from India (where, soon after AD 600, Brahmagupta applied algebra to astronomy) and even China.

Out of all this the Arabs themselves began to shape a new mathematics, though mostly by polishing-up the best of other men's ideas and even then proceeding in a rather untidy way. For hundreds of years there were few advances to which a name or a date can be given: then, in the seventh century, the prophet Mohammed detonated a cultural explosion and the desert folk swept out to carry their faith – and with it, if a little later, their mathematics – as far east as China and northwards into Europe. This latter penetration took the form of a great pincer movement, the eastern thrust of which eventually reached to the gates of Venice and Vienna while the western one pressed far up into France.

It was to be nearly a thousand years before the eastern advance was fully contained. In the west the Arabs had only a few centuries of dominance over such Spanish cities as

Cordova. But even there, though their religion soon retreated, their mathematics remained.

This mathematics, which still bears the typically-Arabic name of algebra, was not an entity in the way that Greek geometry was: on the contrary, it had emerged almost unconsciously from its roots in the Babylonian dawn and did not acquire a full vocabulary until the seventeenth century. Nor was its basis so systematic and logical. The language of this new mathematics, indeed, evolved rather as the local languages of Europe themselves did, and to trace out its development would demand a book in itself. We can here give only a few pages to the birth of algebra; and so, even more than when surveying Greek geometry, we must make use of perfected ideas to explain earlier stages. It should be remembered, though, that by compressing so much we are slightly distorting too. For in the first thousand years of algebra, even more than in the history of mathematics as a whole, language and ideas were very closely associated.

We have, in fact, already met two of the key concepts behind all algebra. The first is that of the generalised number, symbolised by the writing of letters or arbitrary signs to show that we are stating universal rules. The second idea is that of the equation: this, of course, is common to geometry and pure arithmetic as well as algebra, and many of the proofs earlier quoted here have involved the concept of establishing equalities and the techniques of combining them with, and so transforming them into, other equalities. If $A = B$, for instance, we seem safe in saying that $2A = 2B$, that $A-1 = B-1$, that if $A+C = B+D$ then $C = D$, and so on, whatever A and B represent.

The use of equations, though, is not confined to dealing with the universal abstractions which appealed to the Greeks. An equation such as $A+A = 2A$ is really only a statement of two ways of expressing the same idea and should properly be written $A+A \equiv 2A$ – the trebled 'equals' sign showing that this is what is more correctly called an *identity*. Mathematical language is slack on such points; but to appreciate the work of the Arabs we should remember that it was their mathematicians who first systematically used equations in a second way, to identify particular but (for a while) unknown numbers. The very word 'algebra', in fact, seems to mean something like 'the art of solving equations'.

Ali owns a certain number of sheep. If he had as many again, and then two more, he would have as many as his brother Abdul, who has a flock of twenty-two. How many sheep does Ali own? This kind of problem, which dates right back to the manuscript of Ahmes, remains familiar in those books on elementary algebra which live in a world of grandfathers who have forgotten their ages, drivers who do not know the speed of their cars and boys who wonder how many coins they have in their pockets; and it is solved by the device of representing (for example) the number of Ali's sheep by an arbitrary symbol – say X – and then going ahead as if this were in fact known. Thus, we can summarise the information given above as: $X+X+2 = 22$. By performing the same operations (of subtracting 2 and then halving) on both sides of the equation, we discover that $X = 10$ – a fact which we then have to translate back from the world of symbols to that of sheep and shepherds.

There are two important points to note here, of which the first is that the kind of mathematical manipulation we have used in this trivial search for a particular number obeys exactly the same rules – adding like to like produces like, and so on – as the Greeks formulated in their apparently more important work on universal truths. Secondly, the problem itself *is* trivial and little more than a riddle, for in reality nobody ever reckoned-up sheep in this way. (The typical Hindu problem, however, was even wilder and concerned wizards flying off mountain-tops in search of beautiful brown maidens with fluttering eye-lids.) But such brain-teasers appealed to Bedouin shepherds killing lonely night-watches and to the soldiers of the armies of Mohammed as they sat about their camp-fires.

So – not for the last time in history – a great branch of mathematics arose out of a popular pastime. It was to be awaiting when the world needed its methods to calculate the *genuinely* unknown quantities which cropped up in the investigation of the physical world. But the only practical use which the Arabs themselves had for at least their more advanced calculations was in the astronomy which itself served both navigation and the pseudo-science of astrology.

As we have mentioned, there is a fascinating story behind the development over a thousand years or more of the language of algebra. But we shall appreciate the Arab achievement more clearly if we straight away introduce (for those not familiar with them) some essential ideas of this notation.

First, then, algebra uses letters to stand for both generalised and unknown numbers: there is no firm rule, and in its search for symbols modern mathematics raids non-Roman alphabets and even invents shapes, but there is a tendency to use capitals and letters from the beginning of the alphabet for generalisations and for numbers fixed so far as a given problem goes, and small letters from the end of the alphabet for unknowns or 'variables'. Since x is so commonly employed we must avoid confusing it with the multiplication sign of a cross; and in fact the latter is rarely used in pure algebra, where it is enough simply to write two or more symbols together (as in Ax) to express their product.

Another valuable innovation was the use of brackets to

group the 'terms' in an equation which were to be handled as a whole whenever such use made things clearer: thus, $\surd(x^3 - 7)$ means the square root of $x^3 - 7$, and (as the reader can check) is different from $\surd x^3 - 7$. Similarly, the results of the problem in adding areas which we presented on p. 31 can be neatly expressed as: $a(b+c) \equiv ab+ac$.

The only other algebraic form we need to introduce at the moment offers an example of the way in which, in mathematics even more than in the sciences, ideas and language interact. We have seen how the Greeks envisaged the multiplying of a number by itself as the formation of a physical square, with a second such multiplication leading to a cube. But a further multiplying-up did not lead to any mental picture; and so, even in their arithmetic, the Greeks showed little interest in 'powers' of a number above that of the cube. Though they despised the evidence of their senses, they were curiously limited by it.

In algebraic notation, however, we can write aa as a^2 and aaa as a^3, the small figure – known as an 'index' or 'exponent' – showing the number of times a quantity is to be taken before the multiplying-up. Now, if we start discovering interesting facts about the second and third powers of numbers (or, for that matter, about square and cube roots) there is no reason for us not to go on investigating equations involving x^4 or $\sqrt[5]{y}$ (which latter symbolises the fifth root of y), even if we cannot picture what these symbols mean. A trivial-seeming improvement in the expression in writing of a mathematical idea such as this has often led to an important liberation of thought.

But as late as the seventeenth century mathematicians were alarmed at the idea of multiplying four lengths together; and it should be stressed that none of the above ideas were carried into their present form by the Arabs, who simply took further steps along a road first trodden by the Chinese 2,000 years before. One Arabic method of referring to an unknown quantity, for instance, was to call it a 'heap' or 'pile' and symbolise it as a small pyramid – which was at least nearer to the modern method than the Hindu way of referring to unknowns as 'the red', 'the blue', and so on. Yet even with such limited equipment, the mathematicians like Hovarezmi who flourished in the great schools of Persia, north Africa and Spain (and especially in the Baghdad of Haroun-al-Raschid and his son) during the ninth century were able to carry algebraic methods far beyond the handling of equations which involved only the first power of one unknown.

There were, for instance, expressions involving x^2. A simple 'quadratic' statement such as $x^2 = 10$ is easy enough to solve even if the answer turns out to be irrational. But we can construct equations (even when nature does not present us with them) which involve multiples of x *as well as* of x^2; and in the 'general form' in which mathematicians are always most interested the quadratic equation appears as: $Ax^2+Bx+C = O$. The multipliers A and B, it may be mentioned here, are called the 'coefficients' of x^2 and x respectively: C, a simple number, is a 'constant' term.

The general solution of this type of equation is one of the hurdles which the second-year student of algebra has to overcome; and it involves so much manipulation that he probably overlooks the ingenuity of the trick at the heart of the method. (Basically, this is to take advantage of the fact that, if the product of two numbers is zero, then one of them must itself be zero). But the Arabs were not content to lay down a procedure: following the work of Hero of Alexandria they were also able to produce a forerunner of that formula which still has to be memorised by every young mathematician and which expresses the solution of any quadratic in terms of A, B and C. With this, a page of calculation can be replaced by a few seconds of mechanical work.

Similar progress was made in handling the systems of 'simultaneous' equations in which there was more than one unknown quantity. For example, the equation $y = 3x$ is by itself 'indeterminate' or insoluble; but if we know that at the same time $y = x+8$ then we can arrive at the unique solution $x = 4$. The Arabs realised that in all such cases the solver needed as many distinct equations as there were unknowns – and again showed the way towards general solutions in terms of the coefficients.

In their attempts to solve various types of equation the later algebraists acquired considerable skill in handling symbols – and, perhaps, began not to worry too much what these symbols stood for. Suppose, for example, that they were faced with the problem of squaring the two-part (or binomial) 'expression' $A+B$. They found that this could be done just as an ordinary long multiplication sum could if care was taken to separate the various powers of A and B; as it must be to separate units from tens and hundreds in number-calculating.

$$\begin{array}{r} A+B \\ A+B \\ \hline A^2+AB \qquad \\ AB+B^2 \\ \hline A^2+2AB+B^2 \end{array}$$

The working is shown here; and that the result is true of ordinary numbers can be checked (though not *proved*) by putting any values we like equal to A and B. The diagram overleaf provides a geometrical illustration of the identity $(A + B)^2 \equiv A^2+2AB+B^2$ such as the Greeks used; but the algebraic attack on all such problems is much more widely applicable and 'powerful'.

Yet another important by-product of the Arab interest in equations was that this forced them to regard not merely fractional but negative numbers as useful nuts-and-bolts in the

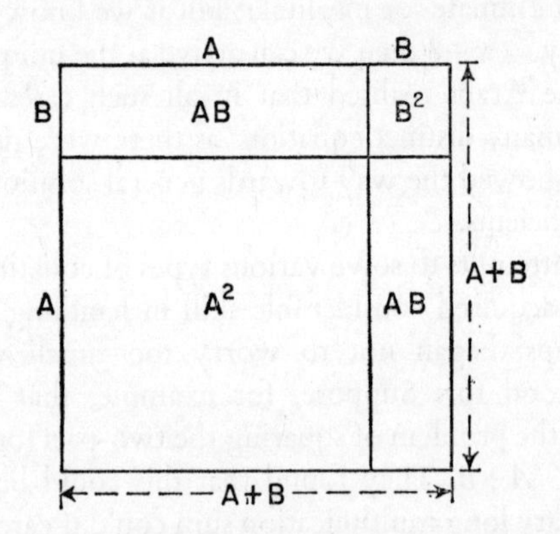

mathematician's tool-kit. Even the innocent-seeming first-degree equation can take several shapes, and of these only one – typically, $x-5=0$ – leads to an answer ($x=5$) in terms of the 'natural', God-given positive integers. If we instead take as our starting point $5x-1=0$ we are led through the fraction-producing operation of division to the answer $x=\frac{1}{5}$. And if we begin with $x+5=0$, the operation of subtraction takes us into the world of negative numbers such as -5.

It is typical of the difference between two mentalities that, whereas the Greeks had found it hard to accommodate negative numbers to their sense of what was fitting, the more empirical Arabs were more prepared to deal with them as mathematical objects without worrying too much about what they represented. On these terms, they found that the negatives obeyed rules which were self-consistent (which is the first test of all mathematical validity) but, at first sight, rather surprising.

The subtraction of a negative number, for example, turned out to be equivalent to the addition of a positive one, and the product of two negative numbers was also positive. If these rules seem unnatural, that is mainly so because of another weakness in ordinary mathematical notation – the use of the '–' symbol to mean both the *operation* of subtraction and the *result* of subtracting a large number from a smaller one. It would have been better if, from the start, some quite different sign had been adopted for the negatives: about this period, for instance, the Chinese were writing them in a different coloured ink from positive numbers. But as it is we must accustom ourselves to the facts that not only does $-(-1) = +1$ but $-1 \times -1 = +1$ too. The former calculation we can at least envisage in terms of repaying a debt, but multiplication by a negative number has few equivalents in the 'real' world.

When we go on to look at equations of the second (and higher) degrees, we meet two other types of number. An equation of the type $x^2 - 5 = 0$ leads to our old friends the irrationals – but what about $x^2 + 5 = 0$? Again the starting-point looks innocuous enough. But the Alexandrians had discovered that when we try to solve this simple equation we find nothing in our repertory of numbers so far to be adequate.

What we are being asked is to find a number which, multiplied by itself, will produce -5. The answer is certainly *not* $-\sqrt{5}$; for we have just seen that the product of any two negative numbers is a positive one. $-\sqrt{5}$ is, in fact, an alternative square root of $+5$, just as important mathematically (and, in some cases, as a solution of practical problems too) as is $+\sqrt{5}$.

F

One of the great conjectures of the Arabs was that, in principle, every equation had as many solutions or roots as the highest power of the unknown quantity represented in it. (Thus, $x^5 = 0$ has five roots – though in this case they all equal zero!) But this fact, which in any case was not clearly proved before 1800, did not help in expressing those square roots of negative numbers which cropped up in equations without any hint that the mathematician had a right simply to disregard them: what seemed needed was yet another type of 'unnatural number', somewhere between the positives and the negatives. Mathematicians, though, had just taken one great leap in the dark by accepting the negatives, and for the moment they were ready to dismiss $\sqrt{-5}$ and its relations by the rather question-begging device of calling them 'imaginary numbers'. After all (they argued, reverting to the Greek habit of visualisation), one could not picture a square of negative area.

This discussion of the number system may seem to belong in the remoter air of the philosophy of mathematics, though we shall see its importance later. By contrast, the final achievement of the Arabs which we must consider at this stage concerned a matter of such practical importance that it has been difficult to postpone mention of it until its proper moment in history. For it implies the greatest advance in mathematical (and not just algebraic) notation that man has ever made.

Fifty or so pages, and almost as many centuries, have gone by since we saw the Babylonians groping towards a counting system of the modern type – i.e. one where the position of each digit indicated whether it stood only for itself or was to be multiplied by a power of the chosen radix. What the Babylonians lacked, though, was a 'place-holder' to use whenever a representative of a certain power was missing from the series. Without this they could not construct a fully positional system, one in which the same symbol could indicate without ambiguity 3 or 300,000.

Although the Greeks and Romans had come closer to modern notation in one respect by using a consistent radix of 10 (rather than the Babylonian 60) to group their whole numbers, they had lost sight of the much more important idea that position could help to indicate value. After a period when a radix of 5 had been in favour (Homer's word for 'to count', for instance, was 'to five it up'), the Greeks had simply added a few letters to their alphabet, split this into three groups of nine, and then made each letter stand for from one to nine units, tens or hundreds: when they moved into the thousands they had to employ modified letters. The Roman system, with its mixture of several traditions of notation, is still horribly familiar from its use on monuments.

But a few centuries after the birth of Christ (it seems) the Hindu school made its greatest contribution to world mathematics by realising that, if a simple place-holder was introduced into the number system, then every integer and rational fraction could be expressed by using a range containing only as many characters as the size of the radix demanded. This place-holder – nowadays written 0, and also known in the mysterious civilization of the Mayas – would in any case have been needed by later algebraists to express possible equation-solutions, to allow for generalisations, and to drop in as shown

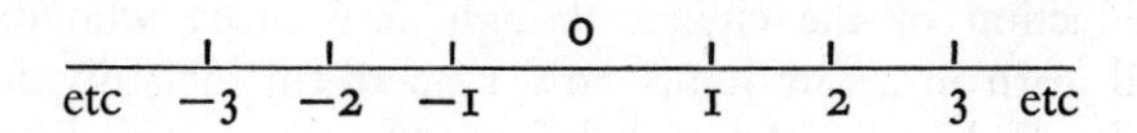

to complete the 'number line' once the negatives had been given official status and the idea of 'direction' in a number vaguely appreciated. It is true that the use of a zero brings troubles of its own to algebra, whose rules have to be specially framed to take account of such facts as that *division* by zero yields a meaningless result. But by making possible the systematic expression of any number as a chain of digits (e.g. one thousand and eight equals: $1 \times 1000 + 0 \times 100 + 0 \times 10 + 1 \times 8 = 1008$), its introduction enormously aided simple calculating in every branch of mathematics. In time, the decimal system was to extend a similar notation to fractions, so that four one-thousandths equal $0 + \frac{0}{10} + \frac{0}{100} + \frac{4}{1000}$, or ·004.

Until the introduction of the 'cipher' zero, in fact, the word 'simple' can hardly be applied to *any* calculations. For the Greeks and Romans had only primitive multiplication tables, and regarded (say) the division of 159 by 17 as an operation to be performed by a specialist – though, perhaps, a merchant-banker rather than a philosopher-mathematician. If this sounds like an exaggeration, the reader can try dividing CLIX by XVII *without using modern ideas* – and then go on to an attempt to extract the square root of the former number.

The more we put ourselves inside the minds of the men of the ancient world, indeed, the more astonishing do their mathematical achievements appear. Even day-to-day figuring was painfully laborious, and – to use a surgical metaphor again – the delicate work of Euclid's predecessors on the irrationals and the infinite resembles that of those Stone Age men who trepanned brain tumours with flint knives. With the introduction of the cipher, though, any smart schoolboy could learn in a few hours how to perform long divisions which would have demanded from Plato's most dedicated pupil the use of elaborate 'short-cut' tricks memorised over years.

The physical sciences were in later years to learn important facts by studying vacuums. But they never learned anything as important as resulted from this *mathematical* 'invention of nothing'; and so it is surprising that Roman figuring (and, with it, the need to carry out calculations with the help of the 'abacus' type of counting-frame today exiled to the kindergarten) lingered on in Europe as late as the sixteenth century. In the far east our numerals have still not taken over completely, and hence the 'swampan' remains popular there.

The cipher was, as we have seen, an invention of the Hindu people who had then taken the lead in the study of numbers. Just when the Arabs became aware of its power, and added it to the now-familiar 1, 2, 3, etc., to produce the series still miscalled 'Arabic' numerals, is uncertain; but it was probably well before the end of the eighth century and the date of the oldest surviving western manuscript to employ a zero. Even after that the growth of the system was slow, though it received a boost from one of the great if half-legendary figures of an age of darkness. For the carrying of Arabic numeration from Spain into the rest of Europe is attributed to Gerbert of Aurillac, the shepherd-boy who preached the first crusade and who is also credited with inventions ranging from gold-plating to the weight-driven clock.

Under the name of Sylvester II, Gerbert became the first Frenchman to be created Pope; and his office was attended with a special significance since he occupied the throne of St Peter in the year AD 1000. In this year many Christians expected the world to end. Instead, though, Christendom was about to break out into a new splendour; and so this is a fitting date at which to turn our attention to Europe north of the Pyrenees.

The five hundred years or more which had followed the ending of the ancient Roman empire are not unjustly called the Dark Ages. The darkness was not absolute, and wherever we find patches of light we find too a groping towards an understanding of mathematics. But this knowledge seems very limited and unprogressive when we compare the Christian to the Islamic achievement.

Recovery came sooner than was formerly recognised, though, and even before AD 1000 it is possible to sense in the air of Europe the first hints of that immensely complex rebirth called the renaissance. If only through the importance of astronomy to the deep-sea navigation which had remained a lively art in the age of the Norse raiders, mathematics then became almost the first non-religious study to capture the imagination of the West.

At first sight the eleventh and early twelfth centuries still belong to the Moslem school; for in this period, when important summaries of mathematical knowledge were also being made in India, two distinguished astronomer-mathematicians were at work at opposite extremes of the Mohammedan world. One of these was perhaps the last great Arab philosopher to work in Spain – Gebir-ben-Aflah, whose name is *not* incorporated in the word 'algebra' and who was *not* Geber the early chemist. The other lived in Persia and is there revered mainly for his work in gathering together a host of scattered results to form a consistent theory of equations, though in the English-speaking world he is best known under the name of Omar Khayyám as the inspirer of a famous nineteenth-century poem. Victorian poetry has also immortalised a leading Jewish mathematician of this period, the rabbi Abraham ben Ezra whose travels once took him to London.

But all this time the mathematical work of the Arabs was percolating into the mind of Christian Europe. Little by little, too, the classical manuscripts which had been guarded in the libraries of Constantinople became known, until scholars had fairly direct access to the minds of Pythagoras, Euclid and

Archimedes. A European student working around AD 1150 had hence two mathematics to study – the Greek, with its accent on the geometrical, the discrete, the precise and the formally logical, and the Arabic which was more flexible, more attuned to changing and uncertain quantities, and perhaps more abstract. These approaches did not always blend happily, and indeed their confrontation by Greek standards of proof was to prove almost a death-blow for the more pragmatic Arab scholars.

In the West, by contrast, the new learning gathered strength. For instance, at Palermo in Sicily, not very far from Archimedes' home town, a Christian university eclipsing even the greatest of the Moslem schools was founded early in the thirteenth century by one of the outstanding figures in the history of ideas – Frederick II, the 'world's marvel', the agnostic ruler of the so-called Holy Roman Empire and the man who (in addition to many half-magical achievements) is believed to have made the use of Arabic numerals official throughout his territories. Soon Bologna, Paris, Oxford, Cambridge and other cities also had universities where a scholar might study mathematics for its own sake. And even earlier the first mathematician of this new Europe could have been seen at work in Italy – the widely-travelled merchant Leonardo of Pisa, otherwise known as Fibonacci, whose keen mind was applied to such fundamental problems as the relation between the discrete and the continuous around the year 1200.

And yet, for over three centuries, mathematical progress seems slow, irregular, and – above all – lacking in great names and consistent themes. In other *Men and Discoveries* books we can see how the sciences fell under the same shadow, and that there a major brake on progress was the opposition of the Church to an approach to problems based on experiment rather than on speculation from tightly-defined dogma. In mathematics, though, there was no such restraining force. For mathematical thinking and education had the blessing of those sages of the classical world who were revered by the Christian Church as giants who had lived in a golden age.

Such 'virtuous pagans' as Pythagoras, indeed, were regarded almost as supermen, and were carved together with saints and angels and holy emperors on those great cathedrals which still witness to the glory of Europe in the twelfth and thirteenth centuries. Even mathematics itself formed an honoured part of the highly-formalised curriculum of medieval education: its *trivium* and *quadrivium* may have neglected algebra along with the sciences, but they included arithmetic and geometry as main subjects even when the arithmetic was largely figuring and the geometry mostly a mindless re-hash of Euclid. There was also a mathematical element – with Pythagoras presiding – in the rules of Church music, and astronomy retained its importance.

But outside the work of Nicole Oresme – a brilliant Norman bishop, mathematician and economist whom we shall meet later – there was little of that rethinking of mathematics which the age demanded; and after the intellectual vigour of the earlier Middle Ages had declined into stagnation amid wars and plagues towards the end of the fourteenth century, new achievements came about mainly by extending the methods of the Arab algebraists in an *ad hoc* way to meet the needs of the technologies of the time. For from 1200 to as late as 1700, practical Europeans looked to mathematics mainly for help in three different, if closely-linked, directions.

Perhaps the most unexpected of these was finance. As men began to engage in trading ventures on a scale not known even in ancient Rome, merchants as well as royal houses had need of large sums of capital. To borrow these sums interest had to be paid, as it had been 3,500 years before in Babylon. And this could be done in several different ways.

Assume that Lombardo is lending a duke £100, and that the unfortunate aristocrat agrees to an interest rate of 100 per cent. Under a *simple* interest system, of course, this means that D pays L £100 every year until he can return the £100 'principal'. But if he repays neither the principal nor the interest at the end of the first year, he will then owe £200. Almost certainly the Lombard will insist that the whole of this sum should itself be subject to interest at £100 per cent, so that after two years the debt will be £400. After three it will amount to £800; and so on.

This is the principle of *compound* interest, and the formula for working out D's indebtedness after a given number of years is fairly obvious. But suppose that L decides on the apparently minor change of making-up his books every six months instead of annually. He then decides that after the first half-year D owes £150, so that his interest for the second half should properly be £75 and the total debt at the end of the year £225. Since this seems a profitable line of thought, the money-lender calculates what would happen if he worked things out over three months. The total debt at the end of the year, he finds, would then be about £244.

In fact, if 100 per cent interest is charged n times a year the debt at its end will be the principal multiplied by $(1+\frac{1}{n})^n$. But this does not mean that there is no limit to the sum which L can claim to be reasonably due to him, for this expression does not increase steadily as n increases. On the contrary, just as the *series* $1+\frac{1}{2}+\frac{1}{4}+\frac{1}{8}$ . . . never adds up to more than 2 however many terms we take, so $\left(1+\frac{1}{n}\right)^n$ itself *converges* to a limit as n becomes indefinitely large or 'approaches infinity'.

The actual value of this limit was soon realised to be a matter of great importance – and not only in the commercial transactions where it also entered into calculations regarding annuities, life assurance and so on. For the extreme case we have just considered is that of a debt growing, not by artificial jumps, but continuously *and at a rate always proportional to its size.*

Now, money is certainly not the only thing which can grow: in fact, during the later medieval and early renaissance periods, it was regarded by a Christian Church which had apparently failed to study the parable of the talents as rather scandalous that riches should be put to work to make greater riches. And the kind of mathematics we are considering was in time found to be applicable to all kinds of problems of growth and decay in the physical and even biological worlds. The $\left(1+\frac{1}{n}\right)^n$ ratio, for instance, could be calculated from the shape of a snail's shell, the rate of a human 'population explosion', the speed of a chemical reaction or the spacing of the rings in a tree-trunk; and one of the curiosities of mathematics is to be found in the series of integers: 0, 1, 1, 2, 3, 5, 8, 13, 21... in which each term is the sum of the two previous ones. For this series, which was first investigated by Fibonacci, is not only equivalent to a discrete or 'stepped' version of the type of growth we have been discussing (the ratio between successive terms soon narrowing-down towards an interesting constant), but is itself represented in nature by the spacing of leaves on certain plants.

The same kind of mathematics crops up when we approach problems of which the Greek search for a 'golden rectangle' is the simplest case: the proportion between the sides of such a rectangle, in fact, is the ratio given by the Fibonacci series. Towards the end of the period we are discussing, too, the musicians who were Bach's ancestors were grappling with such problems as that, if one moved from a 'fundamental' note to an exact 'harmonic' of it and then took the same step upward from that, the overtone *of* the overtone would not be in precise harmony with the fundamental. The only solution here was to tune instruments to a compromise, 'equal-tempered' scale in which every note stood in the same relation to the one below it. And to do this required an understanding of the mathematics of increase.

Even so simple a problem of sub-division as choosing the best lengths of nails or grades of gravel to make up a range leads back to the 'limit' of $\left(1+\frac{1}{n}\right)^n$; and eventually mathematicians came to realise that this was a number of such unusual interest and importance as to justify the honour of being given the special symbol *e*. On investigation it turned out to be irrational and expressible only as the sum of an infinite series; and it was one of the outstanding achievements of a period which was learning to tame such endless chains of numbers to show that the fractional part of *e* could be expressed in the interesting shape: $1/2+1/2\times3+1/2\times3\times4\ldots$ This comes to rather less than $\frac{3}{4}$, *e* itself equalling approximately 2·7182 . . . .

We shall meet this magic number again and again; for it lies at the heart of much mathematics, and not only of that most obviously connected with regular change. The two other technological advances of this period were connected with the more obvious kind of change which we call motion. Firstly, the invention of gunpowder and its use in artillery in the Hundred Years' War had set men thinking more deeply than in bow-and-arrow days about such problems of *dynamics* (or bodies in movement) as why a projectile reached its greatest range when fired upwards at an angle of about 45°; and later the years around 1500 became those of the great sea voyages which opened new continents to European eyes.

As Prince Henry the Navigator had appreciated when he set up the world's first technical college in Portugal, deep-sea sailing laid its own demands on mathematics. Some such problems were not to be solved for more than 200 years more. But navigation was closely linked to the art of map-making (and hence to land surveys of the Egyptian type) on the one hand, and to astronomy on the other.

We have already seen that position-finding had at its heart the idea of *triangulation*, of discovering distances by calculations from observed angles. To take a nautical example of an idea just as applicable to surveying the earth or the skies, suppose

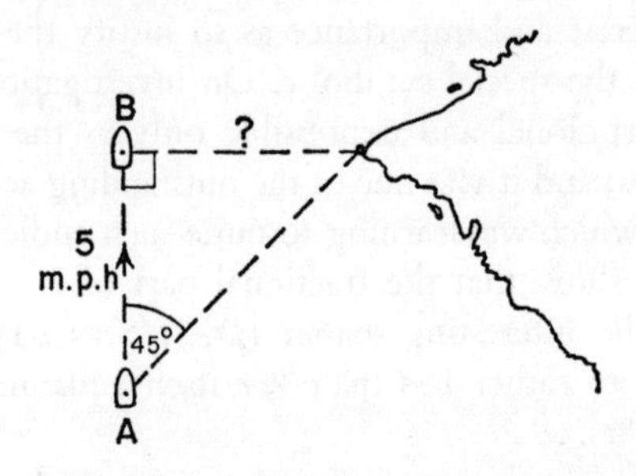

that a ship sailing at 5 mph notes a headland 'bearing 45° starboard' at position A in the diagram. An hour later, at position B, its navigator observes the same promontory exactly on his 'starboard beam'. How far off-shore is he now?

In this very simple example it is easy to see that at B the ship is 5 miles from land. But we must look for a more general attack. It is more likely, for instance, that our first bearing would have been something like $49\frac{1}{4}°$, and our second one perhaps $73\frac{1}{3}°$. We still have adequate information, for Euclid showed that a triangle can be drawn in only one way given one side and two angles of it. Furthermore, of course, we can always construct a scale diagram (making

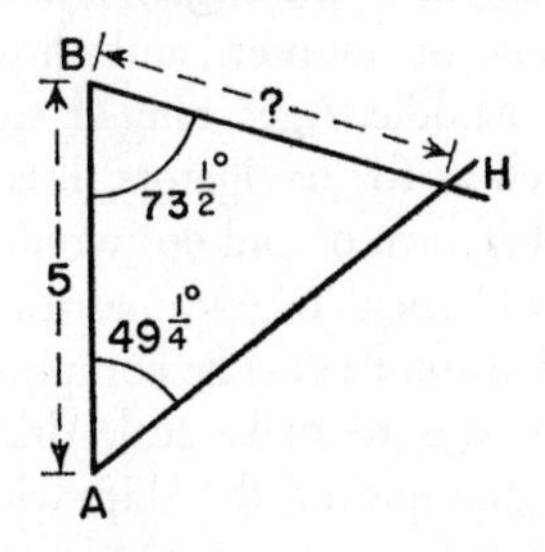

perhaps one inch represent each mile) and simply *measure* BH. But a method based entirely on calculation will be far more convenient and accurate.

Although such Greeks as Hipparchus and the great geographer Ptolemy had made considerable progress along these lines, it was the Hindus who first realised how a whole system could be built up by considering the right-angled triangles into which all others could be divided. In the triangle shown

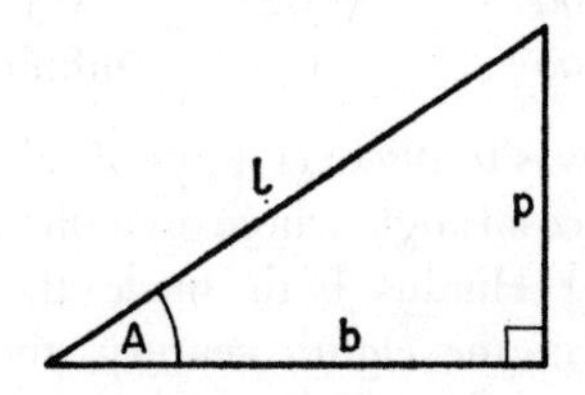

here a 90° angle is marked, and we can also label one angle A and identify the three sides as base, perpendicular height and 'long' side. Now, the Greeks knew that (at least on a plane surface) the *ratios* of these three sides to each other depend only on the angle Â. And so for every value of Â there is a characteristic ratio for p/l (first called the *sine*, meaning 'curvaceous bosom', of the angle in the twelfth century), for p/b (the so-called *tangent* of the angle), and so on.

For land-surveyors, astronomers, and a host of other people who even in the Middle Ages ranged from architects to artillerymen, as well as for navigators, lists of the sines and tangents of angles between 0° and 90° would be of enormous help. But how could such 'trigonometrical' tables (and *this* term was first used around 1515) be compiled? We have seen that a crude attack was to make scale drawings. But even earlier, from our example of the ship with its convenient bearings of 45° and 90°, we saw that at least some such ratios could be calculated. Simple geometry, in fact, gives us this miniature table of trigonometrical ratios, from which it is obvious that sines and tangents are not directly proportional to their angles.

Angle	*Sine*	*Tangent*
0°	0	0
30°	$1/2$	$1/\sqrt{3}$
45°	$1/\sqrt{2}$	1
60°	$\sqrt{3}/2$	$\sqrt{3}$
90°	1	infinite

The next step was to discover ways of calculating (say) the sine of half or twice an angle whose own sine was known. The Alexandrians and Hindus both made their contributions here; then, late in the eighth century, the knowledge of trigonometry passed from India to the Arabs who drew up full and accurate tables. But it was left to Europeans working centuries further on to take the most important step of all and to discover methods by which the trigonometrical ratios of *any* angle could be computed algebraically. Three famous names here are those of the German from Königsberg who (following the custom of the mid-fifteenth century) latinised his name to Regiomantanus, of the Austrian called Rheticus who later spent twelve years to produce tables of an accuracy never known before, and of the Frenchman Viète. Viète or Vieta, a brilliant amateur who reawoke interest in series of all kinds in such time as he could spare from cracking enemy codes for King Henry of Navarre, is also credited with being the first mathematician to use a *consistent* notation of letters in his algebra.

The systemisation of plane trigonometry was another achievement, brought about by the needs of technology, which went beyond the general state of the mathematics at its time as known to such men as Scipio Ferro. Equally important (though this problem too had been faced by the astronomers of Alexandria) was that progress was made with the problems posed by the fact that the surface of the earth was *not* a plane and that Euclid's geometry was hence not valid for long-distance navigation. But quite apart from the great usefulness of trigonometry was the interest of the discovery that – for instance – 'sine Â' could only be expressed in general terms as an infinite series – or in a form equivalent to such a series, that of the 'continued fractions', also known to the Alexandrian Greeks, which take such shapes as this:

$$\cfrac{1}{1+\cfrac{1}{1+\cfrac{1}{1+\cfrac{1}{1+1\ldots}}}}$$

This particular example, incidentally, leads to another expression of the 'golden rectangle' ratio.

Furthermore, such trigonometrical series were very similar in form to those which summed-up to *e*. And, finally, angles themselves could be expressed as the ratio between the radius of a circle and the length of the arc of its circumference swept out by the angle, so that 2π radians would represent a complete turning round of 360°. All this suggested a link between not merely the natures but the actual values of two numbers which, though they could never themselves be precisely expressed and were not even commensurable with each other, seemed to lie at the heart of at least one aspect of mathematics.

Surprising as the relation between the apparently geometrical π and the apparently algebraic *e* is, however, it is little stranger than the fact that the three worlds of finance, navigation and gunnery, all so closely-knit historically and so important in forming the modern world, should also prove to have a mathematics in common. The gunnery, we should mention here, owed much to the work carried out around 1530 by Niccolo Fontana, the son of an Italian postman. Fontana had no reason to love war, for at the age of twelve he had been left for dead in a massacre. His mother nursed him back to life, though, and soon he was teaching himself mathematics, using tombstones to write on because he could not afford even a slate. But his palate had been damaged, so that all his life he spoke with the stammer which gave him his nickname of Tartaglia.

This new mathematics certainly used the methods of ordinary algebra, but its subject-matter was rather specialised. Hard though it is to define, it was concerned with change and motion rather than with fixed (if unknown) quantities; and its methods involved the handling of series which converged to limits which – though definite enough – were not only irrationals but irrationals of a peculiar type. For though

$\sqrt{2}$ could not be expressed as a fraction it was at least the root of the equation $x^2 = 2$, whereas π and e seemed even less tangible than this.

Soon, in fact, the study of such series and numbers was to be treated as a branch of mathematics almost as distinct as was trigonometry from the ordinary algebra of equation-solving – which was itself brilliantly codified at the close of this era by England's first outstanding mathematician, Thomas Harriot, the Oxford graduate who surveyed Queen Elizabeth I's American colonies and brought back to England the comforts of tobacco. But – as we have seen in other books in this series – for every new sub-division of a science there is a fresh bridge built between existing ones. Shortly we shall be examining the greatest of all such bridges; but first we must look at another mathematical world which was opened up by the key called *e*.

4

Giants of the Renaissance

Over a third of this book has passed without our having yet reached the seventeenth century. But although perhaps a hundred times more mathematical knowledge has been gathered since 1600 than in all earlier ages, this does not mean that this account is seriously out of balance. For in mathematics – as opposed to the sciences – the work of pioneers in the medieval, classical and even earlier periods lies not on the fringes of the study but close to the heart of it.

The greatest advance, though, had come quite recently. Consider a man who lived at the end of the sixteenth century returning to life today. If he had learned all the mathematics which had been then discovered, he would probably be able to sail through a modern 'O-level' test paper and even make a reasonable attempt at 'A' levels in some branches of mathematics. Certainly he would understand most of the symbolism used by his examiners, as his predecessors a few centuries earlier would not have done.

That this is so is largely a tribute to the importance of printing. The use of type, of course, transformed all European learning in the century after 1500. But its effect was perhaps greater in mathematics than in any other field, and for a profounder reason than that in this period textbooks on 'commercial' arithmetic became so widely distributed that the merchants of Europe no longer had to use counters for their reckonings and at last became used to Arabic numerals.

The deeper importance of printing was connected with the freedom brought to mathematics by the use of a clear and universally-accepted language. In 1500, for instance, Europeans were still setting out their arguments largely in words, such as 'A number, and 3 times that number *quadratus*, less 7 . . .': a century later, after the work of a group of pioneers such as Stevin for whom we have too little space here, most were expressing the same idea in such forms as $x+3x^2-7$. To mention just one advance, the $=$ sign was introduced in 1557 by an English mathematician in a book titled *The Whetstone of Witte*. And a good notation not only makes the mathematician's task enormously simpler (it is hard, for instance, to imagine how the pioneers carried out with the means available certain operations on series which can give trouble to a sixth-form specialist today), but suggests important new ideas too. Here, at least, there is some truth in the claim that 'the medium is the message'.

Seventeenth-century mathematics was itself dominated by three great advances, of which the third stemmed from the second and all demanded new symbolic languages. In the first of these cases, though, the notation was not entirely novel. For we have seen how, well before 1600, men were expressing the idea of n values of x multiplied together as x^n, so that – for example – $a^3 = aaa$ and $5^4 = 5\times5\times5\times5 = 625$. In all such expressions (and the second example shows that one virtue of this notation is that it allows large numbers to be expressed in a compact form), the symbol in the 'x', 'a' or '5' place is known as the *base* and that typified here by 'n', '3' or '4' as the *exponent* or index.

Now, suppose that we are faced with the problem of multiplying together two powers of x – say, the third and the fourth powers, x^3 and x^4. As Archimedes realised, the product of a group of three xs and a group of four is a group of seven, x^7; and so a multiplication is transformed into a simple addition. The rule can be confirmed by choosing examples from the table of powers of 2 given here: for instance, $2^3 \times 2^5$ *does* equal 2^8. And its inverse is that divisions can be carried out by the subtraction of exponents, so that $2^6/2^2 = 2^4$.

$$2^2-4$$
$$2^3-8$$
$$2^4-16$$
$$2^5-32$$
$$2^6-64$$
$$2^7-128$$
$$2^8-256$$
$$2^9-512$$

Another rule which follows from the idea of bases and powers is that we can 'extract' a square root by halving an exponent: thus, $\sqrt{2^8} = 2^4$ and $\sqrt{2^4} = 2^2$. (Similarly, the *cube* root of 2^9, $\sqrt[3]{2^9}$, will be found to equal the $2^{9/3}$, or 2^3, which logic tells us it must do). But if we can give a meaning to $\sqrt{2^8}$ and $\sqrt{2^6}$, can we not to $\sqrt{2^7}$? The fact that the logical answer of $2^{3 \cdot 5}$ (a group of three-and-a-half 2s multiplied together) cannot be envisaged is perhaps unfortunate; but so is the fact that we often cannot envisage what $-x$ means. Yet by going on despite that earlier difficulty a consistent and useful mathematics had been derived.

Mathematicians do not like having – for example – one rule for even numbers and another for odd ones unless there is a better reason than prejudice for so doing; and hence several algebraists between AD 1200 and 1500 took a leap in the dark. (The outstanding name here is that of Oresme.) Such men *assumed* that the rules that $a^m/a^n = a^{m-n}$, and that $\sqrt[n]{a^m} = a^{m/n}$, applied even when they led to inexplicable exponents. And so, once again, a useful notation led to the invention of a broader and more general mathematics than that from which it had sprung.

For instance, what happens when we investigate $\sqrt{a^2}$? This is by definition equal to a itself, so by our rules a^1 must equal a. This seems sensible enough, but what about a°? We can get at this by seeing that $a^0 = a^{1-1} = a/a = 1$: i.e. any number to the 'noughth' power equals unity, and *not* the zero we might casually expect. Furthermore, since $a^1 = a$, $a^{1/2}$ must be that number which, multiplied by itself, yields a – i.e. $\sqrt{a}$. And finally, if we divide a lower power of a by a higher one we get equations such as $a^2/a^3 = a^{2-3} = a^{-1}$: since a^2/a^3 is also equal – by the simple algebra of cancellations – to $1/a$, then $a^{-1} = 1/a$. (Similarly, $a^{-2} = 1/a^2$, $a^{-3} = 1/a^3$, and so on.) We can hence give meanings to not only fractional but negative exponents: the former turn out to represent roots, and the latter reciprocals.

The previous page contains some of the more tightly-packed mathematics in this book; but the reader who is worried at the building of such remarkable meanings on a mere symbolic rule (for what on earth does a fractional exponent, let alone a negative one, imply in terms of our first ideas about powers?) can perhaps set his mind at rest by looking at the actual values of some of the lower powers of 2. We can here introduce the symbol $\simeq$ as meaning 'approximately equal to'.

$2^2 = 4$; $2^1 = 2$: so what value shall we give to the intermediate $2^{1\cdot5}$? The answer, from our rules, is that $2^{1\cdot5} = \sqrt{2^3} = \sqrt{8} \simeq 2\cdot82$. $2^{0\cdot5}$, of course, equals $\sqrt{2}$ ($\simeq 1\cdot41$). 2^0, as we have seen to our surprise, equals 1; $2^{-0\cdot5}$ is $1/\sqrt{2}$ ($\simeq 0\cdot71$); $2^{-1} = 1/2 = 0\cdot5$; $2^{-1\cdot5} = 1/2^{1\cdot5} \simeq 0\cdot35$; and $2^{-2} = 1/2^2 = 0\cdot25$.

Now – considering only such exponents chosen at intervals of 0·5 – we can draw up a table. What may strike us is that the values follow each other *smoothly* – not, of course, in direct proportion, but at steadily-increasing intervals.

2^{-2} —0·25
$2^{-1.5}$—0·35
2^{-1} —0·5
$2^{-0.5}$—0·71
2^{0} —1
$2^{+0.5}$—1·41
2^{+1} —2
$2^{+1.5}$—2·82
2^{+2} —4

Furthermore, if we calculate a few other results such as $2^{0.333}$ (the cube root of 2, about 1·28) we will find that they too fall into line. Using the idea of a graph introduced a few pages farther on, we could indeed draw a smooth curve linking y to x in $y = 2^x$; and this fact may incline us to trust our mathematical logic.

Even had these ideas stood on their own we would have had to include them in this book. But they turned out to lie behind the most important advance ever made in the logistic or calculating applications of mathematics. So long as men could give a meaning only to positive integral exponents, the rule of $a^m \times a^n = a^{m+n}$, though it replaced the task of multiplication by the simpler one of addition, was little more than a curiosity. But as soon as the ideas we have been discussing were fully appreciated new possibilities opened up.

What we have seen is that any three numbers can be linked by an expression in the form $N = n^m$: given two out of N, n and m, we can always calculate the third. So, as soon as we have a table based on a fixed (and almost arbitrary) n which gives an m-value for every N-value, we can replace the tedious operations of multiplication and division by additions and subtractions, and also transform really tricky jobs of extracting roots into simple divisions.

This new mathematical tool was the next of those which were to be sharpened in the seventeenth century; and that century itself is noteworthy in two ways. It saw the foundations laid for many of the advances which were to bring about a technological revolution, and it witnessed a shift of European inventiveness towards the north.

Ever since Fibonacci's time the centre of mathematics had been in Italy, with Alpine Germany acting as a satellite. But after the work of Harriot and Vieta the seventeenth century was to belong largely to Britain and France – and so equally to these two countries that one historian has suggested that mathematics was no longer like a circle with a centre but rather an ellipse with symmetrical foci. If we refer to one of these foci as being in 'Britain' rather than England it is because – a century before Scotland's political union with her neighbour – an intellectual flowering had broken out about Edinburgh.

It was on the outskirts of that city that John Napier, a contemporary of Shakespeare's, was born in 1550. An important 'laird', and a student at St Andrew's university who had also studied in the more cultured atmosphere of France, Napier became an inventor specialising in weapons of war – and so impassioned a believer in the theology of John Knox that he considered his most important work was advocating the burning of his fellow-Christians. But his immortality in this world is that he was the inventor of *logarithms*.

If we go back to our expression $N = n^m$, we can say that m is the logarithm of N to the base n; this is expressed as $m = \log_n N$, so that $\log_2 8 = 3$. Napier approached the matter much less directly, but it was he who first drew up a systematic table of logarithms and published this in 1614 after twenty years of laborious and unpaid work; he modestly introduced it with the claim that there was 'nothing (right well-beloved Students of Mathematics) that is so troublesome ... than the multiplications, divisions ... and ... extractions of great numbers, which besides the tedious expense of time are for the most part subject to many slippery errors ....' And such was its obvious usefulness (and its inventor's fame, for there is the classic story of the admiring Henry Briggs making a fortnight-long journey from London simply to *look* at the Scotsman, though he went on to spend a month with him) that very soon after Napier's death three years later practical men as well as mathematicians were regularly employing logarithms. By 1620, for instance, his Danish friend Kepler was using them in his important astronomical calculations.

This fact would have been particularly satisfying to Napier, with his typically Scots preference for the useful side of mathematics. He is also remembered, for instance, for the invention of an ingenious calculating device known as 'Napier's bones'. But his 'log tables' – which also helped to popularise the use of decimal fractions – were a much more versatile innovation.

Thus, if two rulers were marked out proportionately to logarithmic scales and then laid side by side so that they could be slid past each other, distances added would represent numbers multiplied and the product could be directly read off. Such 'slide rules' were introduced in England only a decade or so after the publication of Napier's discovery: today there is hardly a scientist, an engineer or a man in one of a dozen other professions who would not feel lost without one on his desk or in his pocket.

We have left one question so far unanswered, that of the base to be used for compiling our logarithmic tables. For practical purposes, in 'common' logarithms, the base employed is 10 and the logarithms are written as $\log_{10}$: this base has the great advantage that once a table has been drawn up of the logarithms of every number from 1 to 10 it virtually repeats itself over the range 10–100, and so on. But almost as soon as Napier had published his work Briggs and a Swiss mathematician showed that, for use in all pure mathematics, the 'natural' base was a very different one.

It was, in fact, none other than our old friend e: and indeed the series most suited to calculating logarithms were themselves similar to those which led to e, π and the trigonometrical ratios. It is not really surprising, though, that this irrational should appear at the very heart of the system of logarithms. For if we look back at the table on p. 84, we will see that equal *arithmetical* increases in the power of a number (for instance, in the series 2^{-2}, 2^{-1}, 2^0, 2^1, 2^2...) lead to a 'geometrical' or multiplying growth of the result: thus, these expressions equal $\frac{1}{4}$, $\frac{1}{2}$, 1, 2 and 4, each of which terms is twice its predecessor. What we are concerned with in the theory of logarithms, as when we were discussing Lombardo and his loan, is the mathematics of steady growth.

Just as important as the 'noble invention' of logarithms was that of the graph. Graphs and their language are so familiar today that we hardly think of them as being mathematical or even scientific devices at all: the editors of popular newspapers and TV programmes certainly expect their audiences to appreciate the way in which a slanting or curved or jagged line can present in a vivid and easy-to-appreciate fashion how road deaths have increased year by year or the popularity of a government has changed, whilst that typical tool of modern laboratory research, the cathode-ray tube, is essentially a graph-tracing machine. But – surprisingly enough – graphical presentation was not greatly used as a method of displaying data, even in technical literature, much before the present century.

In another form, however, the graph had been part of mathematics since Greek times; for we have seen that the idea of a *locus* was that of a point, moving according to some rule, whose successive positions were shown in the form of a continuous line. It is in fact a comparatively minor matter whether we make that line visible by some mechanical device (using a pencil in a loop of string, for instance, or photographing a meteor's fall on a long-exposure plate), or treat it simply as a concept. Either way, some of the first curves to attract interest when Europe awoke to its new interest in mathematics were those traced out by freely-moving bodies whose loci showed relations between position and time. And though part of this interest arose from the needs of artillerymen and clock-makers, the main incentive towards a new approach to the problems of motion had come from the work of those outstanding astronomers of the sixteenth century who paved the way for Kepler.

The great names here are those of Copernicus, Tycho Brahe and Galileo. The first two were almost purely astronomers; but among his other achievements in physics Galileo had investigated such 'dynamic' problems of movement on earth as the falling of weights and the swinging of pendulums. None of these men who dragged the data of motion down from the heavens, though, were themselves major mathematicians. And as the seventeenth century opened the challenge was to translate such data into laws and formulae.

This implied the analysis of the movements of planets and moons (with, as a by-product, the movements of cannon-balls and pendulums too) in terms of the forces acting on them. First, though, there had to be a system for recording their trajectories and orbits, the loci of their successive positions. All of that treatment of motion and force which was the greatest achievement of applied mathematics in the seventeenth century was hence illustrated by the graph, with its ability to represent continuous change.

The same notation, as we shall soon see, also created a new geometry and had an immense influence on pure mathematics. Inevitably the general idea of it had deep roots and had occurred to several men at almost the same time. But only one of these worked through with the idea systematically, and the only predecessor to whom *he* had to look back was once more Orseme, two centuries earlier.

That man was René Descartes: in English his name is usually pronounced 'Day-car', though the adjective from its latinised form is given four syllables, Car-tes-i-an. The son of a minor nobleman, Descartes was born in Touraine in France in 1596 when the renaissance castles which still adorn that gracious countryside were new-built. His creative life hence links those of Galileo and Newton and overlaps with those of many other of the founders of today's science and thought.

So far, young Descartes was fortunate; but as a child he suffered from ill-health and so (like many men of genius) missed much regular schooling: he was possibly tubercular. When he was eight he joined a Jesuit school; but the kindly priest in charge of it realised that the boy was incapable of a normal day's work and allowed him to lie in bed until noon or later. So René formed the habit of spending his mornings in solitary thought – a habit which was to shape his life and his death.

Much of his childhood thinking concerned the nature of knowledge and that theory of experiment which had been pioneered a generation before by Francis Bacon; for Descartes, one of the first of 'modern' mathematicians, was also to rebel against the medieval learning which he had been taught and to become one of the earliest of 'modern' philosophers as well. But we are not here concerned with that philosophy which is summed up in the famous remark 'I think, therefore I exist' so much as with the mathematics to which Descartes' interest was turned by an older member of the college – Fr: Martin Mersenne, who remained a lifelong friend and whose own work on the theory of numbers was to suggest ideas to mathematicians down to the present day.

Considering his ill-health, it may seem surprising that from soon after he left school until he was thirty-two Descartes followed – with interludes of study – a military career. But in fact he was a freelance volunteer typical of his class and times (another such being the real-life d'Artagnan of the *Three Musketeers*, whom Descartes almost certainly met on his occasional sorties into fashionable Paris) who found that this life offered plenty of leisure and solitude. Paradoxically, he discovered that only soldiering afforded the peace and quiet which he prized – though he was successful enough at it to be offered a generalship.

In these ten years or more, Descartes studied every branch of science which came his way, from weather-lore to insect life; and he tried to assemble all this knowledge, together with some creative mathematics, into a grand plan for the universe. The religious perplexities which then racked Europe did not allow the publication in full of this blend of brilliant ideas and total nonsense. But an extract from it appeared in 1637, bearing a long and confusing title (for Descartes was very poor at putting his ideas into words) but now generally known as *The Method*.

The most important hundred pages of this work contain the germ of the new geometry which we shall soon be looking at, and they sealed Descartes' reputation as a mathematician throughout Europe. After having for some years led the life of a wandering scholar he made his headquarters in peaceful Holland; and there, being too self-centred to have even considered marriage, the ex-soldier lived quietly in various small towns or worked as a private teacher. In particular he was in demand as a tutor for royal families; and so, after an invitation from Charles I of England had to be cancelled when civil war threatened, his reputation came to the ears of Christina of Sweden when she was herself released from the cares of the Thirty Years' War. Then nineteen years old, Christina was already a much-feared queen – vain, selfish, but hungry for knowledge and physically as iron-hard as the frozen soil of her country.

So in the winter of 1649/50 the delicate Descartes, now 54 years old, found himself in Stockholm. The weather was unusually bitter even for that city, with biting snowstorms swirling round the royal palace; but that did not deter the 'ice queen' from a rule of life which involved rising before 5 a.m. and summoning Descartes to lecture to her in an unheated library. The inevitable happened, and Descartes contracted pneumonia. As a Roman Catholic, he called for the last sacraments. And in February, 1650, one of the leading intellects of the new Europe, long nurtured on comfortable but productive mornings in bed, succumbed to a brainless nordic cult of keeping fit.

Now we must see what was at the heart of his 'method', the secret which (Descartes half-suggests) was revealed to him in a dream before the battle of Prague. The basic idea had been envisaged by Apollonius, and is so simple that we use it in everyday life when we want to identify a spot with reference to a landmark: thus, we might say that our house is five miles north of the centre of the city and two miles west of it, or that in a garden treasure-hunt the prize is buried 'three paces up the lawn and four to the left' from the sundial. Similar directions are embodied in the latitude-and-longitude systems of mapping the earth and skies which have roots in

Babylonia and certainly were well-known in Descartes' time. And in fact the importance of navigation in the sixteenth century was so great that many problems were solved in 'spherics' (or the geometry of the globe) before they were tackled in the simpler form relating to a plane surface, though the world-mapping work of Nicolaus Kaufmann or 'Mercator' was still to come.

Descartes himself began with a flat surface. On this he drew two lines or 'axes', at right angles to each other and meeting at

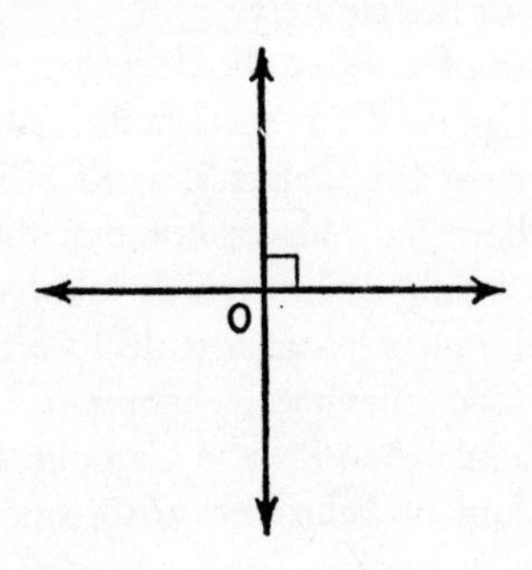

a point called the origin. This is, in essence, all that there is to a 'Cartesian plane'. But it can be improved in two ways – by attaching scales to the axes, and by labelling them according to an agreed convention. We also usually take advantage of the origin to represent zero on both scales, so that negative as well as positive numbers can be accommodated. And so the complete Cartesian grid appears as in the opposite diagram.

Now we have erected a scaffolding. But what are we going to build inside it? In fact, we can construct both a geometry and an algebra. Like Descartes himself, let us begin with the geometry.

The simplest unit in Euclidean geometry is a *point*; and a point can be identified in Cartesian geometry by a pair of values, called *co-ordinates,* showing its distances from the x and y axes. Thus, the co-ordinates of the point A in the figure are 2 and 2, and we can say that B is the point (−4, −3). The next Euclidean idea is that of a straight line;

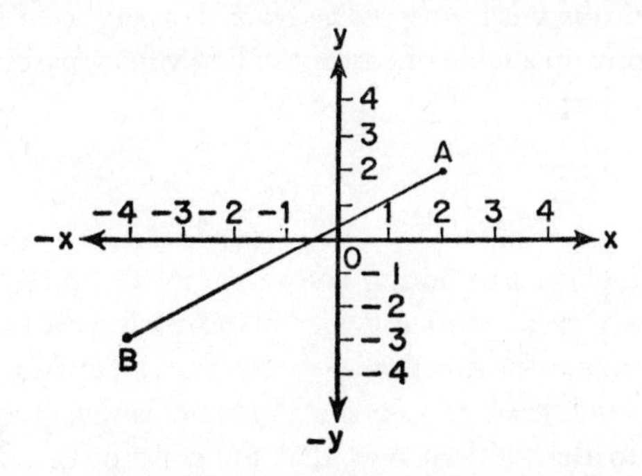

and here our construction of a plane about an arbitrary origin allows us to define the line between these two points as 'the line (−4, −3) (2, 2).' Similarly we can express triangles, rectangles and all the other straight-line figures of Greek geometry in terms of couples of numbers marking their vertices or corners.

To appreciate how this helps us to explore new mathematical fields we must look at the other side of the coin. We have earlier seen how in algebra the 'equals' sign can denote at least two different relationships, the simple identity of – for instance – $x+x = 2x$, and an equation of the type $3x+4 = 10$ which leads to one clear-cut solution (in this case, $x = 2$). But it can also be at the heart of yet another type of relation, one such as is expressed in the 'indeterminate' statement $y = 3x+4$.

We cannot 'solve' this equation in the sense of finding that it leads to any particular values for x and y: there is an infinite range of possibilities. But for every x there is a corresponding value of y, and vice versa: thus, if $x = 5$ then $y = 19$ and if $y = 5$ then $x = \frac{1}{3}$. When a relationship is expressed in this way, y is said to be a *function* of x (the whole statement, $y = 3x+4$, is also known as a function), and x and y are called *variables* with y *depending on* x. For any such function we can draw up a table of corresponding values, part of which looks like this:

x:	-2	-1	0	1	2
$y\,(= 3x+4)$:	-2	1	4	7	10

Four points worth noting before we move on are that we have already used a similar device in examining the behaviour of the 'geometric' function $y = 2^x$, that Descartes was the first mathematician to use an algebraic language virtually identical to the modern one, that the concept of a function has been called the most important idea in all mathematics, and that inherent in it (whatever the logical difficulties

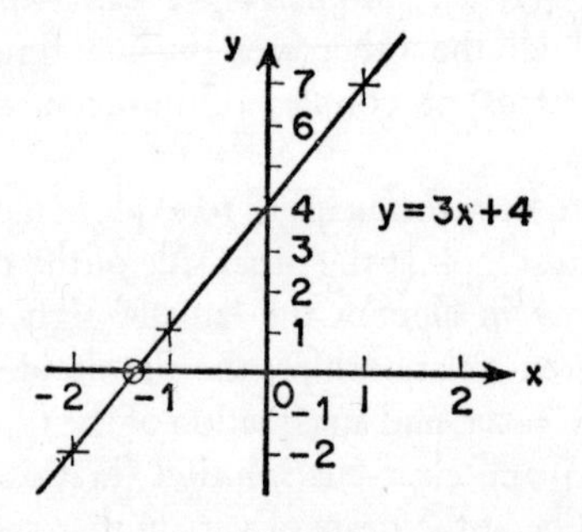

involved) is the concept of *continuity* which insists that numbers like $\sqrt{2}$ and even π fit in naturally on the number-line generated by the rationals.

Now, suppose that we 'plot' all these number-couples on a Cartesian plane, choosing our scales simply for their convenience. The result is shown by the crosses, and at once suggests that the function we have chosen in some way represents (or is represented by) a straight line. We can confirm this by choosing another value for x (for instance, the $x = -1\frac{1}{3}$ marked by the circle), reading-off the corresponding value for y (in this case, $y = 0$) and checking whether this is algebraically correct. After a few such tests – if for no better reason – we are likely to see why a function involving only the first power of a variable (e.g. 3x) and a constant (e.g. 4) is called 'linear'. All such equations, when related pairs of values for x and y are treated as geometrical co-ordinates, lead to straight-line graphs whose slopes are measured by the coefficients of x and whose distances from the origin by the constant terms.

A straight line now appears as a 'curve of the first degree', a 'degenerate' curve. The simplest second-degree *curve* is a circle; and if we analyse the circle in the next figure with Pythagoras' theorem in mind we shall see that for any point on its circumference $x^2+y^2 = r^2$: i.e. the equation of a circle

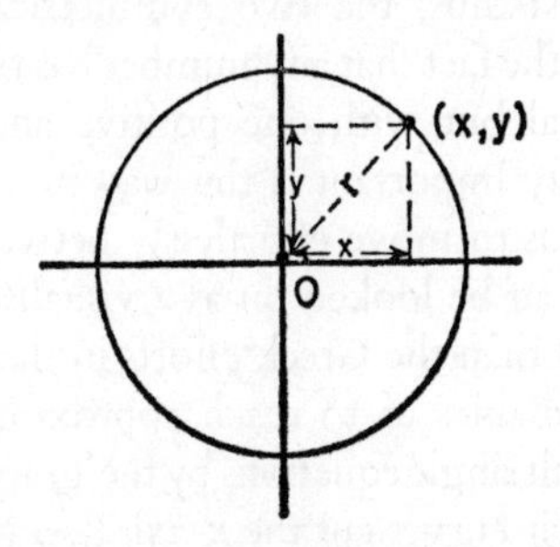

with its centre at the origin is $y = \sqrt{(c-x^2)}$, where c is a constant. For an ellipse, as we might guess, we have to introduce another multiplying coefficient to represent the amount of flattening.

But again we can look at things from the other side and start from the simplest second-degree *equation*, $y = x^2$. What kind of curve does this lead to? When we plot it out (remembering that negative values of x will give positive ones for y), it

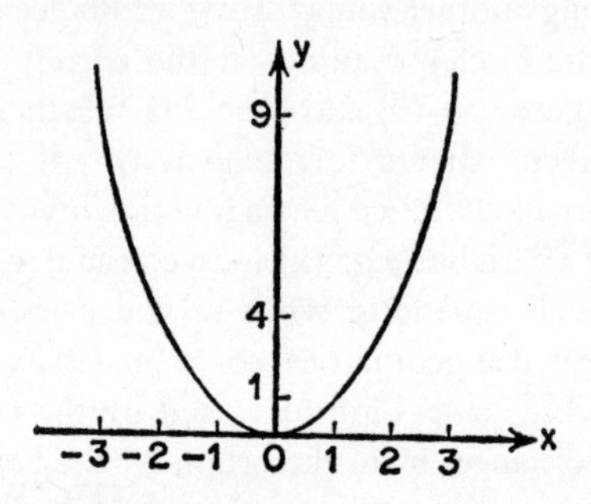

looks like a parabola; and Euclidean geometry shows that the figure indeed obeys the rules which, two chapters ago, we saw to be typical of that curve.

Each of these examples illustrates ideas which could usefully fill several pages: thus, the two symmetrical horns of the parabola display the fact that any number has two square roots, numerically equal but with one positive and one negative. But what is really important is the way in which Descartes' method enables us to move effortlessly between two worlds. For instance, it can be looked on as a visualisation of algebra far more general than the Greek efforts in that direction; and in this guise it enables us to reach approximate solutions of otherwise-difficult single equations by the 'graphical' method of seeing where their curves cut the x-axis (i.e. where $y = 0$), or of systems of simultaneous equations by noting where two curves cross each other.

But since the techniques of algebra were by the seventeenth century becoming very powerful, it is not remarkable that the main effect of Descartes' work was that in future geometry was to an increasing degree looked upon algebraically. Thus, there were theorems about even circles and straight lines which had puzzled the Greeks but which could be solved in a few lines by the use of Cartesian (or 'co-ordinate', or 'analytical') geometry, and its introduction meant that problems about the parabola which had defeated Archimedes came within the reach of anyone who could solve a quadratic equation.

So it is not surprising that, for more than three centuries, the invention of Cartesian geometry has been regarded as one of the greatest of all advances in mathematics. Its techniques,

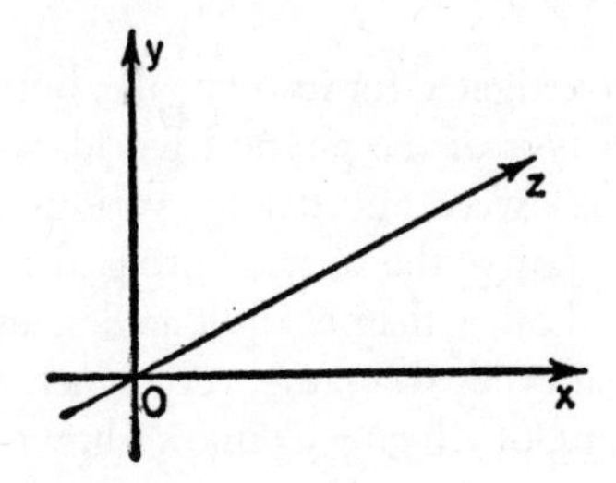

too, were later extended in several ways. For example, by introducing a third axis – shown *in perspective* here as the z direction, or modelled in the round as in Plate 1B – problems in solid geometry could be solved, with the simplest equation for a sphere being (as we might imagine) $x^2+y^2+z^2 = r^2$. This advance came about around 1700.

For special purposes – for instance, in the investigation of spirals as shown overleaf – it was found useful to adapt a different reference system entirely and to use 'polar' co-ordinates in which the position of a point was fixed by its direct distance from the origin and by the angle which the line joining them made with the x-axis; this dates from about 1780.

Another convenient device was to mark-out one or both of a pair of normal axes with equal intervals corresponding not to the 'arithmetical' divisions 1, 2, 3..., etc. but to a 'geometrical' or logarithmic series such as 1, 10, 100, 1000 ...

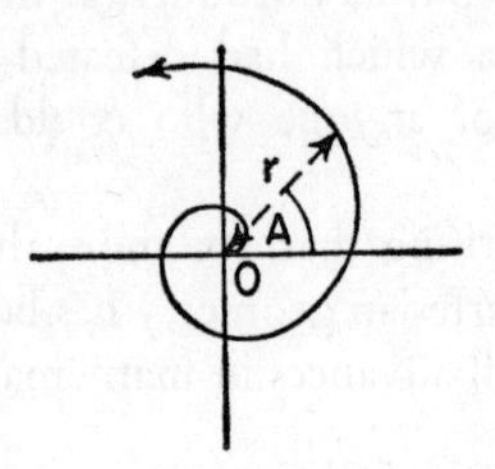

An idea first investigated for its own sake, but later found of interest in the analysis of the physical world, was to see how algebraic equations were affected by various distortions of their graphs – rotating the axes, shifting the origin, using axes which crossed other than at right angles, and so on. And applied mathematics also became very much engaged with 'periodic' functions, of which y = sine x (shown here with the

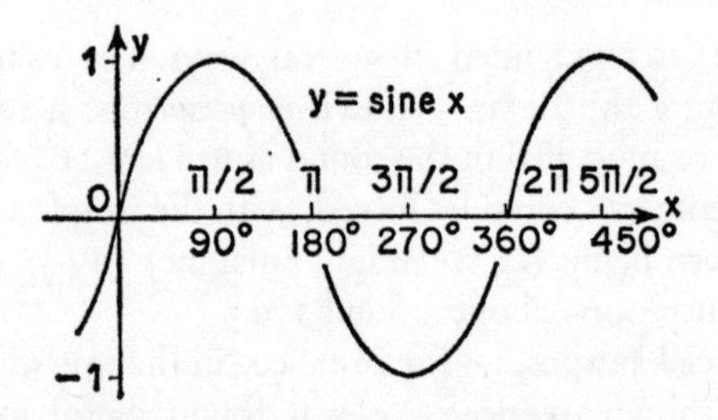

x-axis marked out in both degrees and 'radian' measure) is the simplest and is involved whenever a rotary motion is projected on to a straight line.

Despite this many-sided usefulness, though, the greatest

achievement of Cartesian geometry was probably the freedom it brought to pure mathematics. Some readers, for instance, may have wondered how in the graph above we can introduce *negative* sines, since a sine is the ratio of the lengths of two lines. Yet (as Descartes realised) the new outlook made it reasonable to refer to a negative length, angle or anything else – once one had established a zero to work from.

This fact has perhaps been suspected by the Arabs and even the Babylonians. But the application of Cartesian methods led to light being thrown on an even deeper mystery. This was the nature of the square roots of negative numbers.

As we have seen, these cannot be represented on our ordinary number-scale (there is no part of the curve in the diagram on p. 98 below the x-axis); but they *do* occur in the solution of simple equations. In fact we have really only *one* such troublesome root to worry about – that of -1 itself – since an expression such as $\sqrt{-5}$ can be treated as $\sqrt{5} \times \sqrt{-1}$. Similarly, the fourth, sixth and other 'even' roots of -1 must all be related to its square root. The cube and other odd roots do not present this particular problem, though they too can lead the algebraist into deep waters.

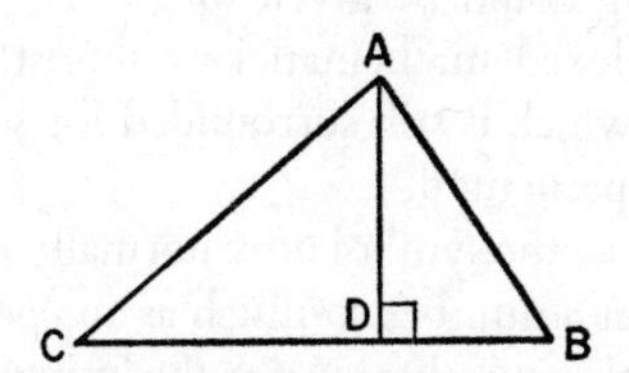

Now, there is a straightforward Euclidean theorem which shows that the square of the perpendicular height of a right-angled triangle equals the multiplied lengths of the two parts into which this perpendicular cuts the base: in the diagram, $AD^2 = CD \times DB$, or $AD = \sqrt{CD \times DB}$. Let us take a triangle simplified so that AD, CD and DB all measure one unit, and impose it on a Cartesian plane as shown in the next diagram. The height of the triangle is, by our theorem, now equal to the square root of the *negative* distance from the origin to the -1 mark multiplied by the *positive* distance from

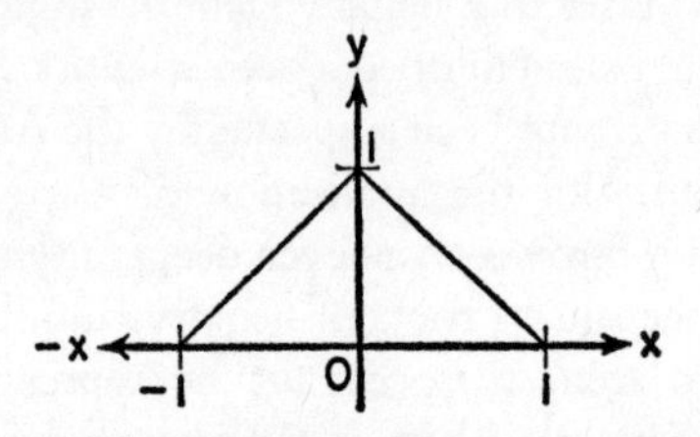

it to the $+1$ point – or to the square root of minus one! The theorem of Pythagoras, which had in Greek times given a meaning to irrational numbers, now appeared to help men to visualise the quantities which Descartes himself had still considered 'imaginary'.

This demonstration (which is not a strict proof) was itself not discovered until the end of the eighteenth century. But long before then it was realised that through Cartesian mathematics a meaning could be given to $\sqrt{-1}$ – that expression which had perplexed mathematicians almost since the birth of algebra, and which is still surrounded for some by an aura of the almost supernatural.

In these terms i – the symbol now normally used for $\sqrt{-1}$ – is regarded as not a number so much as an *operator* which can be attached to other numbers just as the 'minus' sign is. Earlier in this book, we saw how all the 'real' numbers can be regarded as forming a line with a turning-point at the zero mark; we can hence envisage the operation of multiplying by -1 (i.e., of moving from n to $-$n) as the rotation of a number through 180° about an origin. i itself now appears as an instruction to go half-way towards this reversal by 'turning' our number through a right angle – by convention, in the anti-clockwise direction.

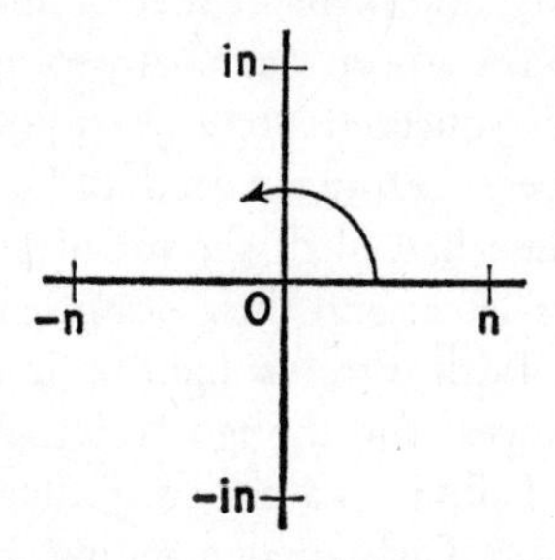

A few checks will show that these rules, like those for the handling of exponents, lead to a self-consistent system. Two multiplications of a number by i, for instance, are equivalent to a multiplication by i^2 or -1: after three such, n is transformed in $-in$; and after four multiplications we are back where we started since $i^4 = (i^2)^2 = (-1)^2 = 1$. But, even so, these ideas appear so remote from our familiar notions of multiplication and the like that many readers may be wondering why we have devoted so much space to them.

Part of the answer is that even in Descartes' time it was realised that the use of i made possible valuable short cuts in such processes as the working-out of trigonometrical series. Later, we shall catch a glimpse of other unexpectedly practical applications of this by-product of his vision. But meanwhile we must look at another development of the graphical idea where it is difficult to say whether the pure or applied aspect has proved the more important. It came about from a break-through made independently by two men; and though such cases of simultaneous discovery are not uncommon, this instance is perhaps the most important in the whole history of ideas.

The two men were Isaac Newton in England and Gottfried Leibniz in Germany, and there are several similarities between their lives. Newton was born – as a baby so small that, he said, he could have been squeezed into a quart pot – on Christmas Day, 1642, eight years before the death of Descartes and a few months after Galileo had died: the infant Leibniz arrived in Leipzig four years later: and their deaths came in 1727 and 1716 respectively. Both were brought up in reasonably comfortable circumstances; and though Newton was to end as a national hero and Leibniz as a half-forgotten man, both were in their time given fairly undemanding jobs by political authorities, with Leibniz acting as a lawyer, librarian and diplomat in the service of the future George I of England and Newton as a senior civil servant and member of parliament.

Both, too, were influenced by the vague and bible-centred mysticism characteristic of seventeenth-century protestantism and very different from the toughly logical theology of the catholic Middle Ages. For Newton this study even became so much of a passion that his mind was wasted for years in searching the book of Daniel for clues as to the end of the world, or speculating on the geography of hell. The German, however, is entitled to be taken seriously as a philosopher, and particularly as a thinker about the nature of mathematics.

By contrast, Leibniz carried out no original scientific work, whereas Newton would be highly regarded for his achievements here alone. (He indeed vastly preferred his 'applied' to his 'pure' investigations – though, like Napier a half-century earlier, he considered his now-forgotten theological ideas superior to either.) It is true that in chemistry and electricity he was a mere alchemistic dabbler, and that even his very important work in optics – and particularly on the theory

In the 12th century, Pythagoras was looked back to as a father of both mathematics and music. Part of the magnificent sculpture of Chartres cathedral shows him plucking at a zither-like instrument

Another kind of mathematical 'sculpture', the three-dimensional graph of a typical function, in this case that of $(x + y^2)^3 + (z - y^{2/3})^2 = A(K - y^2)^3$, where A and K are constants and x, y and z three variables.

Archimedes

Sir Isaac Newton

Karl Friedrich Gauss

Einstein

MATHEMATICAL GIANTS OF FOUR AGES

The development of the calculating machine. Pascal's historic model, part of one of Babbage's ingenious but unwieldy mechanical computers, and a modern IBM electronic computer, which can display the results of its calculations as a graph or reverse the process by producing data from a curve sketched with a light-pen on to a screen.

of colour – was marred by his following of a false theory. But he became a great observational as well as analytical astronomer, and he not only designed but built the first example of the type of telescope, relying on a mirror rather than a lens for the initial stage of its magnification, which is used in all major observatories today. The fact that the type of mathematics which we shall be discussing for the rest of this chapter was at first so closely linked to astronomy hence gives us a reason to look at it mainly through Newton's, rather than Leibniz', eyes.

First, though, we should stress the curious fact that man's first real understanding of the laws of force and movement – his abandonment of the fallacies of Aristotle and his ability to give precise meanings to terms such as 'power', 'energy' and 'work' – came from the study of the skies and not of the more accessible earth. Newton's great laws of motion, too, sketch-out a universe in which bodies not subject to outside forces travel on for ever in straight lines and where feathers fall as fast as bricks – a universe which seems very different from our actual one.

The solution to both these paradoxes is that there were secondary forces, such as friction and air-resistance, which were very important in affecting bodies moving on earth but which scientists could not analyse until they had a clear idea of the *primary* forces involved. Out in space, though, there are virtually no such distractions and we are hence presented with a laboratory where bodies interact in an ideal fashion. Newton may or may not have been inspired by a falling apple; but the planets are a great deal more consistent in their trajectories than apples, and with their year-long cycles easier to watch too.

Newton himself spent a quiet childhood at a Lincolnshire farmhouse, where he amused himself by making toys and models (such as a mouse-driven mill which, 2000 years before, might have amused Archytas) while there raged around Britain the civil war which had broken out in the year of his birth. He also learned some chemistry from a local pharmacist. As yet he had given no hint of unusual talent, but at the age of 18 he won a place at Trinity College, Cambridge. This was just becoming an important mathematical centre; but Newton's studies there were rather haphazard and he was lucky in catching the attention of the university's first professor of the subject, Isaac Barrow.

Barrow, the Master of Trinity, was a man of many gifts and an important mathematician in his own right: he also forms one in a line of thinkers who worked in Oxford and Edinburgh as well as Cambridge and who link Napier's generation to that of the brilliant group which founded the Royal Society. (Of this group Newton himself was to become one of the brightest if most quarrelsome members, while another link-man was the Cambridge graduate and Oxford professor John Wallis, who could not add two and two together before he was fifteen but who became a pioneer of the theory of exponents and of the use of *i*). But Barrow is best remembered for an unselfish gesture. Having seen Newton into a fellowship at Trinity – and a vacancy was not hard to find, since two of the reigning fellows had just fallen downstairs when drunk and a third had been removed to a madhouse – the great professor resigned his chair in favour of his greater pupil and devoted himself to theology. Barrow was under forty at the time, and Newton only twenty-six.

For meanwhile there had been a strange development. The two years which followed Newton's rather undistinguished graduation in 1665 were those of the great plague, and for most of them the university was closed. Newton hence took his work home to his village of Woolsthorpe; and there, in only eighteen months, he not merely discovered his talents at last but carried out the nucleus of his life's work. This nucleus included the forming of at least the first ideas in his theory of universal gravitation – that greatest of all scientific laws, itself based on the meticulous observations of Kepler and his proof that the planets moved in ellipses, which states that any two bodies attract each other with a force proportional directly to their multiplied masses and inversely (or reciprocally) to the distance between them squared. And the bridge between Kepler's observations and this law was itself the new mathematics which Newton founded in a period of creative work which, perhaps, no man has ever exceeded.

With the return of normal conditions, though, Newton became involved with other matters: he seems to have regarded gravitation as almost a trivial affair, and two decades were to pass before – encouraged by the astronomer Halley – he polished-up his work on it for public showing. But then, after another spell of intense creativity, there appeared the book known as the *Principia* – one of the profound achievements of the human mind, and a work which, only fifteen years after its publication in 1687, had revolutionised thinking about not only astronomy but ordinary mechanics. By one of the ironies which characterise Newton's life, his new mathematics as such did not appear in this book, for he considered that to use them there might prejudice the acceptance of his *physical* ideas.

Instead – and this was a mental triumph in itself, though Archimedes had had to do something rather similar – he painstakingly translated the Cartesian arguments which presented a harmonious and comparatively simple picture of the way in which gravitation ruled everything from the orbits of comets to the height of tides into much clumsier proofs framed in the familiar terms of Euclid's geometry. And so we have the best of precedents if we say little more here concerning Newton's physical and astronomical achievements.

When the *Principia* was published its author was forty-five. Forty more years were to pass before Voltaire saw the old bachelor buried in Westminster Abbey with all the honours which the state could offer; but if these years were marked by such public recognition as the first knighthood ever granted to a scientist, they were years of anti-climax too. For, scientifically as well as in pure mathematics, Newton did virtually nothing in the second half of his life except refine and reissue his earlier work.

Yet there is at least one anecdote to show that, had he wished, Newton could have gone on creating brilliant mathematics for at least the earlier of these years. In 1696 Leibniz and a continental colleague had followed a custom of the times and announced a mathematical competition. In this case the puzzle was to find the Cartesian equation of the curve, linking A and B like a ski-jump, down which a ball affected

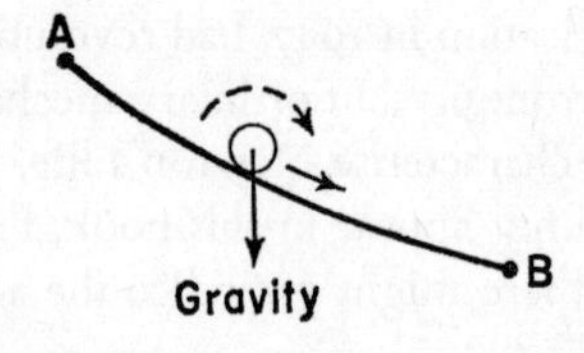

only by the tug of gravity would roll in the shortest possible time. The problem is neither trivial nor simple (for the answer is certainly *not* a straight line), and it had baffled the mathematicians of Europe for six months before Newton heard about it.

One evening he came back from a day of work at the Royal

Mint and had his supper. Unable to get the puzzle out of his mind, he picked up the pen and paper which he always kept at hand and solved it in a few hours. He could not resist sending his answer to Leibniz, but because he hated being regarded as a pure mathematician he did so anonymously. It was no use, however, for the German took one glance and said (in effect) that one could recognise Newton's work as one recognised a lion from its paw-marks.

And now we must look at that great machine which the man who despised mathematics did so much to create, and begin by giving it its modern name of 'the calculus'. The term '*a* calculus' implies simply any system of mathematics – more limited than a whole kingdom such as algebra, but not as limited as the 'tricks of the trade' called algorisms – adapted to producing results; and though the word is not at all systematically used we shall meet other calculuses in this book. But *the* calculus almost always means the method of Newton and Leibniz, which is more properly called the 'infinitesimal' calculus.

This itself has two complementary branches – a fact which adds to its power but makes a further difficulty for anyone who tried to sort out the story of its conception. For the moment, though, we can simplify things by leaving aside its enormously-wide practical applications and instead consider some questions in Cartesian geometry. This course is well justified historically, since in the thirty-odd years which elapsed between the publication of Descartes' 'method' and Newton's own use of the calculus for the analysis of physical problems such questions had engaged the attention of almost every important mathematician in Europe. Among those who had pioneered the attack followed by both Newton and Leibniz, for instance, were the Italian Jesuit, Cavalieri, the map-makers like Mercator who had their own need for means of systematically straightening-out curves, the astronomer Kepler, John Wallis, and Isaac Barrow.

Before we begin, however, we need one more definition. We have already met the word 'tangent' in its rather artificial and confusing trigonometrical sense of a *ratio*: we now need to use it with its original, Euclidean meaning of a straight line which touches but does not cut a curved one. A good way of getting used to tangents is to slide a ruler over a surface of varying curvature such as an egg – though here we need consider only one, cross-sectional plane of such a 'solid of revolution'.

Experiment will soon convince us of the most important fact about tangents, that there is usually only one of them at any given point. We shall also see how a group of them

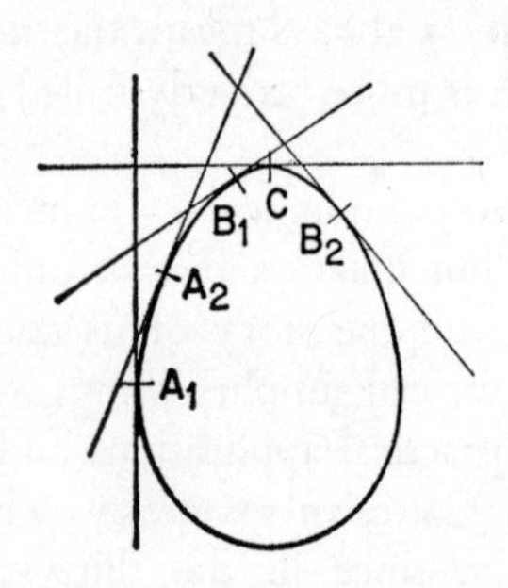

forms an 'envelope' ghosting-out the curve itself, as the five in the figure are beginning to do. And finally we shall discover something very relevant to the ideas of the calculus – that the angle of a moving tangent changes most rapidly where the curve or surface is most sharply bent. For instance, our ruler might swing through only 25° as we moved its point of contact half an inch along the egg from A_1 to A_2, but through 80° in the same distance between B_1 and B_2.

Now let us consider any curve drawn on a Cartesian plane without gaps, crossings or violent changes of direction. We will assume that we can give an equation to this curve, but for

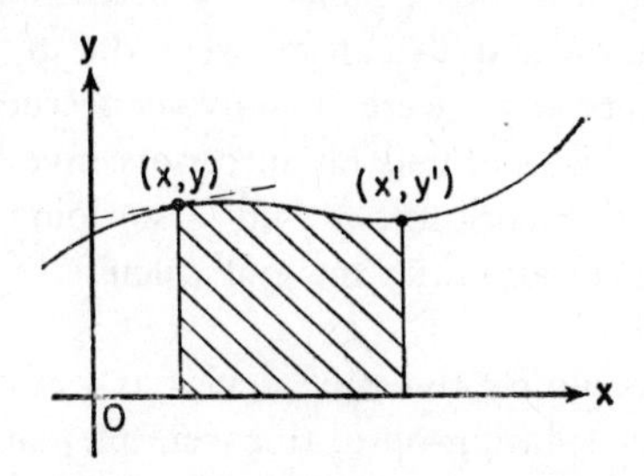

the moment leave open the question of what the equation is to be. Thus, it might involve any numbers of powers of *x*, trigonometrical ratios, logarithms, constants, and so on. All we demand is that every *y*-value shall be a given *function* of a corresponding *x*-value.

Anybody looking at the figure above with pure mathematical curiosity in mind may be tempted towards asking three questions. These are:

(1) What is the slope or gradient of a tangent (the dotted line) drawn to the curve at a given point – say, at (x, y)?

(2) What is the distance, as measured along the curve itself, between two points such as (x, y) and (x′, y′)? (The ticks over the letters here, like the suffixed numbers in the tangents-to-an-egg illustration, are simply a device for calling attention to the fact that the symbols stand in a 'family' relationship to each other.)

(3) What is 'the area under the curve', in the sense that the

shaded zone denotes the area which it encloses by progressing from (x, y) to (x', y')? The units throughout of course, are of only *relative* importance and will be fixed by the scales chosen for the two axes.

If the reader feels he has met something like questions (2) and (3) before, then he is quite right; for the Greeks pitted their brains against such problems of determining both the lengths of curves and the areas swept out by them. (The Euclideans, certainly, were mainly concerned with the circle, but Archimedes and his successors investigated conic sections and other curiosities.) And in so doing they looked ahead to the methods of the 'integral' calculus of Newton and Leibniz.

For this uses almost the same device as was employed by the Greeks, the splitting-up of (for example) an area into an ever-increasing number of rectangles, themselves of ever-decreasing size, whose total area can be summed-up by deducing and then applying some rule. The reason why the renaissance men were able to go so far beyond the Greeks was that in the algebraic geometry of Descartes they had a means of expressing the shapes of curved figures far more general and more powerful than their forerunners knew. But when Newton himself said that he saw further than other men only by standing on the shoulders of giants, he certainly reckoned Archimedes as well as Descartes among his predecessors.

This, then, is all that there is in principle to 'the calculus': geometrically speaking, it is a way of reducing curved-sided figures to straight-sided ones by considering the curves as series of minute straight lines. We have already introduced the misleading adjective 'integral' for the branch of it which deals with lengths and areas. The other branch, which is concerned with slopes, is known as the 'differential' calculus; and it too was mapped-out by the great Archimedes when he showed how the angle of a tangent to a spiral could be calculated.

From this field of mathematics we must now take just one example to represent its method: our argument will follow that used by Leibniz rather than Newton. The curve in the diagram here is the graph of the simple but very important function: $Y = X^2$, or at least of the half of this parabola which

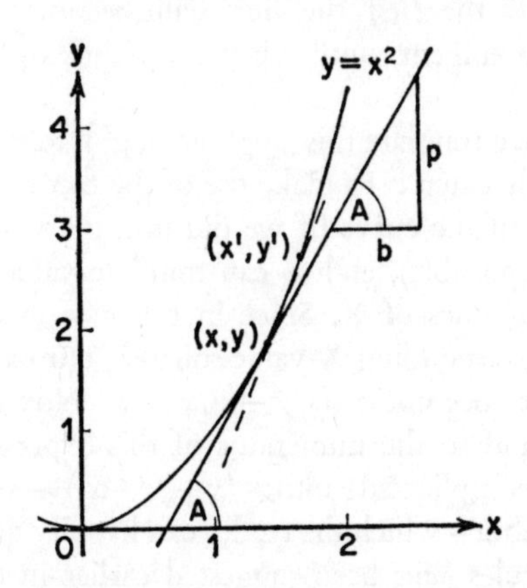

involves positive values of x. What we will look for is a formula to connect the slope of the tangent to this curve at a given point (x, y) with the value of X itself there. (Note that we are using capital letters to denote generalised values and small letters for particular ones.) We *could* define this slope by the angle Â which it makes with the x-axis or base-line; but it is more convenient to use the ratio p/b in a typical triangle drawn with its shorter sides parallel to the axes. This ratio is, of course, the *trigonometrical* tangent of Â; and so it becomes clear how a word which originally meant a certain type of line came also to imply a ratio, often measuring a gradient.

First we consider a point near to (x, y) which we call (x′, y′): it does not really matter which side of the original point this is. A line – here shown dotted – drawn through the two points will make its own angle with the X-axis, the angle whose trigonometrical tangent-ratio is $y'-y/x'-x$: but this line is not itself quite the geometrical tangent *at* (x, y). On the other hand, as we move the second point closer and closer to the first the line will become nearer and nearer to the tangent, until when the points collide it *is* the tangent.

How can we translate this 'approaching' into mathematical terms? The first step is to make use of the fact that we know the equation of the curve (if we did not, the whole exercise would be impossible), and so can translate values of Y into corresponding ones of X. Since in this case every Y-value equals the corresponding X-value squared, our expression for the gradient becomes: $(x')^2-x^2/x'-x$. Now x' equals $x+(x'-x)$, and so the numerator of this expression can be apparently complicated into: $(x+(x'-x))^2-x^2$. A little ordinary algebra – which the reader can happily take on trust, though the rules *have* been suggested earlier in this book – tidies this up into: $x^2+2x(x'-x)+(x'-x)^2-x^2$, or more simply $2x\,(x'-x)+(x'-x)^2$.

This is a convenient moment at which to divide the top by the bottom part of the fraction, bearing in mind that this division will turn the numerator into an expression for the fraction as a whole. This denominator, which we have left untouched, is $(x'-x)$, and so the fraction itself now becomes $2x+(x'-x)$.

And now comes the final step – one which is typical of all calculus methods. The points x' and x approach each other until their separation is *infinitesimally* small, and at the 'limiting' moment which we are most interested in this difference vanishes entirely. And so the gradient of the curve $Y = X^2$ at the point marking the value of X as x is itself just 2x.

We can use similar arguments to 'differentiate' a whole host of functions. (An unusual and interesting one is $y = \log_e x$, whose 'derived' function turns out to be $1/x$. Here e appears once again in a mathematics of steady change, and it is also of importance that $y = e^x$ differentiates *into itself*). We have seen too how such methods can be used in integration or the calculation of such things as the areas enclosed by curves. And a discovery due to Leibniz which paid mathematics a great and unexpected bonus was that integration turned out to be simply the inverse of differentiation, so that if – for example – we knew that the area under a curve was given by $A = x^2$ then we could say that the curve itself followed the general form: $y = 2x$. In this particular case the result can be checked by simple geometry.

But, thought-provoking as all this may be, we have as yet given few clues as to why both branches of the infinitesimal calculus became so immensely important in physics and elsewhere. The short answer is that they are concerned with extracting the maximum of information from a graph – or from any function which can be expressed as a graph, as almost all of them can. To return to that key concern of seventeenth-century physicists, the analysis of movement, the straight line overleaf shows how the displacement of a certain body from its starting-point might be found to vary with its time in motion: and since the 'curve' *is* a straight line we can say that the body moved with a constant velocity, in this case of one foot per second. But suppose that the displacement was *not* directly proportional to the time elapsed – i.e. that the velocity itself was ever-changing. Just as the slope of the line in the first case represented an overall speed, so in the second instance the slope of a tangent to it at a particular point would represent the speed at the corresponding moment.

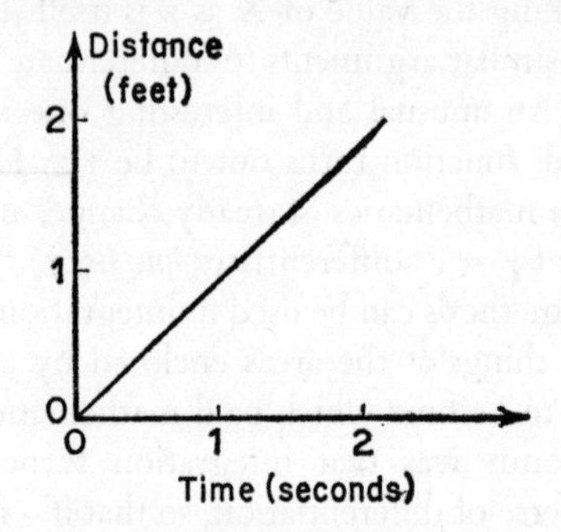

So if – either from experimental measurements or from a study of the forces involved – we can give an equation to such a time/space curve, we have only to differentiate it to find the velocity at a particular instant. In fact, Newton discovered that the displacement of a body accelerated by a steady force such as the pull of gravity would be given by the equation $d = \frac{1}{2} a t^2$, where a is a constant representing the acceleration. As we saw above, $\frac{1}{2} at^2$ differentiates into at, which tells us that the *velocity* of a body moving under such conditions increases steadily with the time elapsed.

Alternatively, had we started from this knowledge we could have worked backward to the original formula by using the integral calculus. To Newton himself, indeed, the great usefulness of both branches was that they provided a perfect instrument for the analysis of the movements of earthly and heavenly bodies alike. For the three ideas at the heart of dynamics are those of displacement, velocity and acceleration; and these are related to each other by successive divisions by time.

Consider, for instance, one fixed body and one free to move. According to Newton's law, the second will fall towards the first under the influence of a force which grows stronger as the distance diminishes. This force produces an acceleration, or speeding-up of a velocity which itself represents distance divided by time. Everything is in a constant state of change; but the changes are ruled by simple laws, and a sure guide through the maze is provided by that infinitesimal calculus which Newton himself strikingly named the method of *fluxions*.

For Newton, though, this was little more than a means to an end – the end of showing why Galileo's cannon-balls moved in parabolas and Kepler's planets in ellipses. As we have seen, his first conception of the infinitesimal calculus came as early as 1665. But he allowed his ideas to be only slowly leaked out by his pupils, and his own account was not published in full until 1736, nine years after his death.

Meanwhile, Leibniz had covered the same ground by a different route. After having been inspired by a meeting with Christiaan Huygens (who, though his first fame is as a rival to Newton in the theory of light, was also a fine mathematician), the German had approached the calculus as a pure mathematical challenge about 1675. And *he* published his ideas less than a decade later.

At the end of the seventeenth and the beginning of the eighteenth century there was hence strong and even bitter controversy as to which man had 'invented the calculus'. Following his spectacular and accurate prediction that a comet would appear in 1680, the future Sir Isaac had become a national hero in the eyes of many who had little idea of what gravity, let alone the calculus, was all about. He had been responsible for something as newsworthy as a moon-landing, and his countrymen were firmly on his side and not slow to point out that Leibniz had had mathematical discussions with Newton before beginning work and that, in some of his enormous range of other activities, he had been less than honest. On the continent, by contrast, Leibniz' supporters looked no further than the fact that his publication had been the earlier by several decades.

There were two especially unhappy by-products of what is now recognised as a case of genuinely independent discovery. The two men most involved never went to the lengths of vituperation of some of their supporters, and remained sincere admirers of each other's work; but the friendship which had existed in the days when Newton had sponsored Leibniz' membership of the Royal Society was at an end. More important for progress, though, was the fact that this severance led to a split extending through the whole world of science and mathematics.

Here the continent lost something by not appreciating for over half a century the full importance of Newton's ideas in astronomy and physics. But Britain lost more in that men of the talents of the algebraist Taylor and the geometer Maelaurin, who was Newton's successor at Cambridge, followed the great man in regarding the calculus mainly as a device for attacking certain physical problems and did not appreciate to the full its role in pure mathematics. It is another of the ironies of the history of ideas that, for nearly two centuries after 1675, Britain became a second-rate nation mathematically largely because its thinkers were too heavily influenced by their brilliant predecessor.

The contrast between the approaches of Newton and Leibniz is reflected even in the notations which they used. Today's method of expressing the ideas of the calculus is almost identical to that of the German, whose main innovation was to use the prefix 'd' to represent an infinitesimally small increase in a quantity symbolised; thus, the slope of a tangent (itself representing the rate of growth of one quantity with respect to another) could be expressed as dy/dx. When an infinite number of such infinitesimals were to be summed up to a finite total the fact was expressed by using as a symbol the old-fashioned elongated s: sometimes called 'summa', this should not be confused with the Greek capital S or 'sigma' which is used when a number of finite quantities are to be added together.

$\int$

$\sum$

The advantages of this notation (on which the fundamental theorem which expresses integration as the inverse of differentiation appears simply as: $\int dx = x$) become important when, for example, mathematicians have to deal with differentials *of* differentials. When we pass from a displacement directly to an acceleration we are dealing with the rate of change of a rate of change (the velocity), a fact illustrated physically by the way in which the slope of our ruler in the tangent-to-an-egg experiment changed at rates varying with the curvature of the surface. Such higher-order differentials are common in the analysis of the physical world; and similarly when engineers have to follow Archimedes and consider the centres of gravity of certain solids, or calculate the bending of beams and the energy in flywheels, they use third and even higher orders of integrals. It is, incidentally, clear from these examples alone that the integral calculus has many uses other than area-finding or the reversal of differentiation.

The uses of the calculus today, indeed, go far outside Newton's world of 'fluxions' and into realms where his own notation becomes almost useless. It would, perhaps, be unfair to take leave of the giant on such a note of apparent criticism. On the other hand, the famous lines which Newton's statue inspired from Wordsworth – and which suggest that his was an untroubled and Olympian mind – present an equally false picture. For like Leibniz (and Wordsworth himself) Newton had his darker side, one which in his case took the form of secretiveness and of delusions that he was persecuted and that anything which he published would be misunderstood.

This man who claimed to see himself as a child gathering pretty pebbles of knowledge on the shores of a boundless sea possessed one of civilisations' towering intellects; but he was also a human being, and sometimes a confused and unhappy one. And though he explored and explained many of the mysteries of space and time, he never unravelled those of his own soul.

5

Coming of Age

The eighteenth century was that in which our subject took up its present form. Hitherto mathematics had been acting largely as a handmaiden to various disconnected technologies, but by 1700 the advances which had been made in many directions were knitting together into a pattern of knowledge. At the same time, the sciences themselves were entering a comparatively brief period during which experiment was to be all-important, so that for a while there was to be less call for new symbolic work. Mathematicians themselves were hence freer to follow their own sense of what would prove important and fruitful.

Before *we* enter this period, though, we must look back for three quarters of a century; for three near-contemporaries and fellow-countrymen of Descartes remain to be mentioned in this book. The reason why we have not met them before is that they were little concerned with that analysis of curves and changes which dominated mathematics from before 1650 until after 1800.

The first of these men was Gerard Desargues, who at the time when Descartes was working out the analytical method developed a new geometry of his own. This – the first important work along Greek lines since the time of Pappus – was concerned with the distortion of lengths and angles brought about by 'projection' – with the way, for instance, in which the shadow of a triangle cast at an oblique angle on to a screen becomes a quite different triangle, or a circle becomes an ellipse. The subject is obviously closely linked to the artist's (and, even more, to the engineering draughtsman's) problems of perspective; indeed, Desargues was himself a military architect and before him both Leonardo da Vinci and Albrecht Dürer had had similar ideas. But it also has a mathematical elegance of its own, which is connected with the fact that for every proof concerning *lines* there appears a corresponding or 'dual' result linking *points*. Hence specialists in this subject nearly always get two theorems for the price of one.

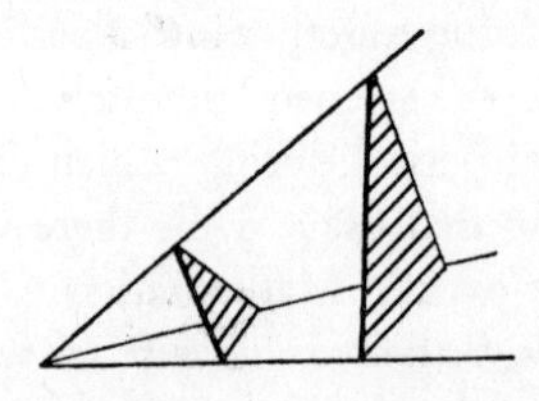

This 'projective' geometry, which Desargues modestly introduced in an article whose title began 'A proposed rough draft of an attempt . . .', also worked usefully as a supplement to Descartes' analytical method. Though the latter eclipsed the former for many decades, for example, Newton showed how the curves of the bothersome cubic equations which he met with in his study of movement could be reduced by projection (i.e. by distortion according to a set formula) to well-understood parabolas.

It is probable that the next great mathematician in order of time is Pierre Fermat – but *only* probable, for there is no certainty as to his date of birth. There is no doubt about the place, however: it was near Toulouse, a city which later employed him as a kind of superior town councillor. Fermat studied as a lawyer there and showed no interest in mathematics until he was thirty; and he remained an amateur – perhaps the greatest in the whole history of the subject –

even after he had been inspired by dipping into the proofs of Diophantus concerning the properties of integers.

For 1,400 years pure arithmetic had made little progress in Europe: it appeared to resemble not so much a backwater as a mountain lake, clear but deep and chilly and only connected by a narrow channel to the main river of algebra. But Father Mersenne had reawakened interest in the theme, and now Fermat was to make a contribution to it second only to that of Diophantus himself.

In the three decades before his death (which occurred in the same year as saw Newton sheltering from the plague in a Lincolnshire garden), Fermat produced dozens of theorems dealing with integers in general and with primes in particular. Many of these are very simply stated, and a few are even as simple to follow as they are elegant. But not one was simple to devise.

With so much ground to cover in this book – and not least in this chapter – we can look at only a single example of Fermat's work here. But this example is easy to choose, for it is not only typical but the subject of the most famous puzzle in the entire history of mathematics. It bears the rather romantic title of 'Fermat's Last Theorem.'

We saw earlier how Diophantus concerned himself with such questions as the possible solutions of an equation such as $A^n + B^n = C^n$, where all the numbers had to be integers. For the 'Pythagorean triads' where $n = 2$ there are an infinite number of solutions; but Diophantus failed to discover a single case for a higher power. Even for $A^3 + B^3 = C^3$ it seemed impossible to find whole number values for A, B and C.

Fermat stated that this *was* impossible and that $A^n + B^n = C^n$ is insoluble in integers for a higher power of n than 2. Furthermore, he claimed to have proved the fact by 'a truly marvellous demonstration'. But all that he wrote on the subject is a brief note in the margin of his copy of Diophantus.

There is nothing surprising in this, for it was the one peculiarity of this busy, quiet and contented man that he rarely *did* write out his proofs in full. (Fermat never published a single 'paper', for instance, all his work being contained in letters to friends.) But everything else which he claimed to have proved has since been confirmed by others, even when it took talented men years to dispose of what Fermat seems to have dealt with in a day. A general proof of Fermat's 'last theorem', however, has defeated the greatest professional mathematicians and the keenest amateurs for over 300 years – and for some of these years there was a substantial cash prize at stake as well as a kind of immortality.

Nobody doubts the truth of the proposition: nobody doubts that the supremely honourable Fermat believed that he had proved it: but he might *just* have been mistaken. Ironically, the theorem itself is less important than many others in the theory of primes which are harder to state but simpler to prove. But some at least of the millions of brain-hours which have been (and still are being) expended on it were not wasted – for, more even than mathematics as a whole, that curious and beautiful corner known as pure arithmetic produces unexpected by-products. And out of the search for a proof of Fermat's Last Theorem has come a mass of important new methods, many of them being of use not only in the theory of numbers but in other (and apparently-unrelated) branches of mathematics.

Fermat is also associated with the development of a very important idea in optics which has its roots in the work of Hero of Alexandria but whose full importance was only appreciated a century ago: this states that the course of a ray of light can be summed-up in the statement that it follows the path which will occupy the least time. But for the most part he was too immersed in the theory of numbers to pursue to their limit the ideas which flowed from his versatile mind.

One of these, for instance, was a realisation of the way in which the curve-studying devices of geometry and the equation-solving devices of algebra could reinforce each other: this would have led to an invention of co-ordinate methods independent of Descartes'. Another such idea was an approach to the calculus thirty years before Newton's. And yet a third was a formation of the ideas of *probability* which were developed by Blaise Pascal.

Born in 1623 at Clermont-Ferrard in the heartland of France, Pascal had a mathematical father who at first discouraged his son from following him but soon became his enthusiastic sponsor. For the boy's genius would not be denied. He gave up all his play-time to the forbidden subject, and before he was twelve had discovered the 'sum of the angles in a triangle' theorem for himself: he went on to produce most of the great Greek proofs before hearing of Euclid and then to similar work on conic sections: and long before he was out of his teens he had made a very important contribution to the new projective geometry of Desargues.

At the age of nineteen Pascal showed a different type of talent by building the world's first workable calculating machine (Plate IVA, see p. 99). This was an adding device of the type in which numbers are clicked up on ten-toothed cog-wheels which 'carry one' to the next wheel in line when each group of units, tens and so on has been completed. The principle was familiar enough; but the difficulties of friction had defeated several would-be inventors after Leonardo of Pisa, and when Pascal was successful the king of France sent for a copy of his machine. It was later improved by Leibniz.

So, well before he came of age, Blaise Pascal was famous.

His climb had certainly been helped by the fact that his family belonged to that influential if intrigue-ridden professional class, characteristic of the seventeenth century in France, which had links reaching up to the court; but his own talents played a greater part. When he was only fourteen, for instance, Pascal had been discovered by Descartes' tutor and Fermat's friend Fr: Mersenne, the mathematician–priest who was now moving on from being a kind of unofficial post-box for scientific correspondence towards the foundation of a French equivalent of Britain's Royal Society.

But – even more than in the case of Newton – there was a darker side to the picture. For no great man in the history of ideas ever suffered from worse physical health than did Blaise Pascal. In his case the childhood sicknesses which afflicted so many talented men were not thrown off as the years passed: instead, Pascal's body degenerated until there seemed no part of it from his teeth to his guts which functioned normally. From the age of seventeen onwards his days were long tunnels of blinding pain and his nights sleepless horrors.

With his super-human brain coupled to a sub-human body, Pascal was doomed to a tortured adult life – though a short one, for he lived for less than forty years. But added to his physical torments were the spiritual ones caused by those bitter theological tangles which we have already seen ravaging northern Europe throughout this century. Pascal chose the worst possible course and joined the heretical sect of the Jansenists who differed from most Romanists and all protestants.

Between about 1645 and 1660, while Pascal's body seemed stretched on a rack, his mind too was torn between mathematics and theology. After an escape from a road accident which the philosopher took to be a symbol of God's will, theology won; and so Pascal, like Newton, put his great work behind him before he was thirty. Because he was a genius, even his *Pensées* remain classics for their literary value in an age when their content has ceased to be of great interest. But one could have wished for more of the mathematical work of which three examples must be mentioned here.

One of these is an isolated curiosity concerning the cycloid curve which we met earlier: this beautiful, useful, versatile and interesting shape gave mathematicians a good deal of trouble before the age of the infinitesimal calculus, but in one brief bout of work which he indulged in only to take his mind off his pains Pascal produced a complete analysis of it. The second field of study, perhaps more physical than mathematical, concerns the perfecting of Robert Boyle's ideas on air pressure and the weights of gases. And the third and greatest is the theory (or, better, the *calculus*) of probability.

The mid-seventeenth century was a great age of gambling in many forms. Some of the reasons for this fact are social, but another is that such problems as whether one was more likely to throw a 6, 4 'main' at dice than a double 5 were imperfectly understood and could be resolved only by a tedious working-out of all the possibilities. Hence a professional gambler who had done a little homework could spend his evenings making an easy living from the less cautious.

One such gamester, more thoughtful if less successful than the rest, ran up against a question of this kind which he could not solve and wrote to Pascal about it. He was lucky to catch the great man in a moment when he was prepared to consider an apparently trivial matter. But it was only *apparently* trivial; and out of this gambler's query was to grow a whole new field of mathematics, something vastly more important than the curses against other Christians to which Pascal (like Napier fifty years earlier) devoted so much energy.

Before taking a glance at this probability calculus, however, we should remember that (like most innovators) Pascal did not begin in a vacuum. Even in Europe the basic ideas of probability had been explored before 1400, and Fermat had just formulated some of its rules. Historians have indeed suggested that Fermat and Pascal would each have done the other's life-work in mathematics had they stuck to that subject – and have deplored the fact that they did not do so.

The claim is probably true, and could even be extended to Napier, Descartes, Newton and Leibniz, any one of whom might have encompassed almost all that was of the first importance in seventeenth-century mathematics had they not 'wasted' their time on theology, philosophy, literature or physics. But we are now describing an age so rich in talent that all the great advances were bound to come within a generation or two, whoever made them. And instead of bewailing a waste of genius we should rather rejoice that the ideal of the all-round intellect, conceived in ancient Greece and reborn in early-renaissance Italy, should have been so fully realised in northern Europe during the seventeenth century.

Now let us return to the two-dice problem mentioned above. At first sight, [6, 4] and [5, 5] seem simply alternative ways of throwing a total of 10 with two dice, each as likely as the other. But in fact there is only *one* way of getting the second fall but *two* of getting the first one, [6, 4] and [4, 6]; and this is hence twice as probable a result. The point may seem obvious today, though plenty of people are still taken in by simple 'catch' questions involving the rules of probability. But it was not at all clear when Pascal began his work, if only because there are considerable logical difficulties involved in defining the terms needed in any fundamental discussion of this branch of mathematics.

There is much less difficulty in recognising its subject-matter, though. It is largely concerned (as in the above example) with the laws of *combination*, with the ways in which objects and events can be grouped together: and from this it moves on to *permutations* or the ways of shuffling order within groups. By building up an ordered structure of possibilities, Pascal founded a study which was to prove of immense value not only in such sciences as genetics, and in 'useful arts' other than bookmaking and completing football coupons, but in the purest of pure mathematics too.

To see how this came about, let us begin with a pastime even simpler than dice-throwing – coin-tossing. A penny can only come down two ways, heads (H) or tails (T), and we can write the (equal) likelihood of the outcomes as [1, 1]. If we toss two pennies – or the same penny twice – there are theoretically four equal possibilities – HH, HT, TH and TT; but for most purposes we can regard the permutations HT and TH as identical, so that our range of *types* of outcome is simplified to [1, 2, 1]. Some of the most intelligent men of the seventeenth century, incidentally, could not see that one was twice as likely to throw a mixed pair as a pair of heads (or of tails).

With three coins or tosses there are eight possibilities in all – HHH, HHT, HTH, THH, HTT, THT, TTH and TTT. (This analysis emphasises that the likelihood of an even chance being repeated n times is $1/2^n$). And thus the chances of types of outcome here (3 heads, 2 heads and 1 tail, 2 tails and 1 head, and 3 tails) are as [1, 3, 3, 1.]

So we can go on – tediously by 'enumerating cases', or more economically and elegantly by spotting the rules which are beginning to appear. The result in either case is a triangle of numbers of which the first eight lines (including a hypothetical 1 to begin with) are shown overleaf. This array was known before AD 1300 to the Chinese, those great gamblers who seem to have had *some* understanding of permutations 1600 years before the lifetime of Christ. But nowadays it is usually called 'Pascal's triangle', if only in honour of the fact that the Frenchman wrote a treatise on its properties.

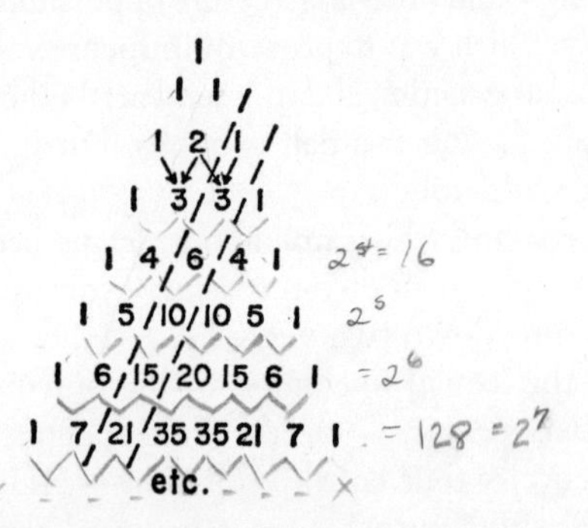

One way in which this triangle can be formed is suggested by the arrows between the third and fourth lines: every number (except for the outside units) is got by adding together the pair flanking it in the previous line. As a mere generator of number-tricks its possibilities seem endless: for instance, the lines – as we have seen – add up to powers of 2, so that the last one illustrated totals 2^8 or 128. The 'diagonals' have their own characteristics (with the one enclosed by dotted lines presenting our old friends the 'triangular' numbers); and only a few years ago an amateur mathematician spotted that even Fibonacci's series – which the reader may like to hunt down for himself – was buried in the array. But our concern now is with the way in which these 'heads or tails' lines enter into algebra.

Earlier in this book we met the term 'binomial' as implying a two-term expression, like $(A+B)$. We also squared $A+B$ to produce the result $A^2+2AB+B^2$; and if we cubed it we would end up with $A^3+3A^2B+3AB^2+B^3$. The coefficients of the various powers of A and B in such 'expansions' are, in fact, the numbers displayed in the corresponding lines of Pascal's triangle, so that if asked to write down the expansion of $(A+B)^4$ we can say without the trouble of multiplication that it will be $A^4+4A^3B+6A^2B^2+4AB^3+B^4$. The analysis of such a calculation (which it is well worth-while the reader's carrying out) shows *why* this should be so: the coefficients represent the number of different ways in which each term arises throughout the long multiplication.

Now, the generalising of the expansion of $(A+B)^n$ – which leads to the important 'binomial theorem' – lay at the heart of algebraic manipulations of the greatest importance to the advance of seventeenth-century mathematics. If we look back at our differentiation of x^2, for instance, we shall see that we had to expand just such a squared term; so obviously if we are to find a general form for the differential of x^n we must first know a general form for the expansion. In fact, Newton produced this for himself – but it was already available as a result of Pascal's work.

The solution of such problems involved the devising of several more new notations, but perhaps the most important innovation of probability mathematics was the use of 'factorial' numbers. Just as a triangular number is the *sum* of a string of successive integers: $1+2+3+4...$, etc., so a factorial is its *product*: $1\times2\times3\times4...$, etc. The modern method of indicating such a product is with an exclamation mark, so that $6! = 6\times5\times4\times3\times2\times1 = 720$, and $n! = n(n-1)(n-2)...\,1$.

It is fairly obvious that we are going to meet such numbers when we consider multi-stage differentiations. But they enter, too, into the handling of a host of infinite series; indeed, we have already met the factorials in our expression of e on p. 75. And so, when one opens a textbook of advanced algebra and finds every other page spattered with 'screamers' or 'shrieks', it does not mean that the author is (like a chess commentator) congratulating himself on his cleverness. He is simply – if indirectly – acknowledging the debt which all mathematics owes to the gambler who inspired Pascal to the study of 'combined' numbers.

The next great mathematician in chronological order would provide a link back into the domain of series if 'he' were a person at all. But in fact 'he' was a family, the most remarkable and ramified one in the whole history of ideas. There are plenty of cases where two or three members of a clan became distinguished in the same or allied fields; and we can leave to the experts the question of how far such genius is hereditary and how much can be explained by family atmosphere. But not even their contemporaries, the musical Bachs, produced in three generations *nine* men of real importance in one specialised world.

Yet this is the distinction of the Swiss family Bernoulli. Most of its cantankerous members worked in the eighteenth century. But the first two brothers, usually referred to in English books as James I and John I, flourished around 1690 and so form a bridge to a new age.

For a moment, though, we must postpone discussion of the work which was to be typical of this period and instead follow a biographical line of descent. The elder of the Bernoulli brothers (who was the first to use $\sqrt{-1}$ constructively, and who had also been Leibniz' colleague in setting the 'falling ball' problem which Newton so triumphantly solved) had taught mathematics to a fellow-citizen of Basle. And when in 1707 this clergyman had a son he was handed over to the younger brother for a similar coaching.

The boy might have gone on to win fame in one of a dozen spheres, for he was another of the infant prodigies in which the history of mathematics is rich – though not discouragingly so. Like many of his period he studied protestant theology: like a good Swiss he mastered languages easily: like any young *savant* of that age before intense specialisation he absorbed all the important physics and astronomy of his time: and he also showed interests in literature, music, medicine and botany. He chose, however, the career which was to lead to Leonard Euler becoming perhaps the greatest mathematician of the eighteenth century and the most productive in the entire history of the subject, the man who worthily united the ages of Newton and Gauss.

The life of Euler (pronounced, roughly, 'oiler') was long and happy, and even his end was enviable: after an afternoon of work on the problems of flight posed by Montgolfier's newly-invented balloons, a dinner with his friends and a game with a grandson, he put down his pipe, said 'I die,' and did so. The only real shadow he ever met was a literal one, for Euler lost one eye at the age of twenty-eight and (after some pain-creating medical bungling) became totally blind whilst he still had seventeen years to live. But the mathematician has less need of his physical senses than almost any other professional man, and Euler's intellect remained undiminished until the very end. Aided by his remarkable memory and powers of mental calculation, he even worked most intensely in his years of darkness.

Euler was particularly happy in his home life, although that was to be established far from his native land. In other books in this series we can see how, in the eighteenth century, centres of research drifted away from the universities of Europe towards royal and noble courts; and another characteristic of the age was that it saw the first of Russia's periodic and energetic attempts to catch up with Western learning and technology. So Peter the Great and his wife

Catherine I combed Europe for men of talent whom they could import, and by 1727 the widowed empress was sponsoring the work of two second-generation Bernoullis in her capital – that splendid new city which was arising on the shores of the Baltic as Alexandria had arisen beside the Mediterranean 2,000 years before, and which was then called St Petersburg.

The Bernoullis told Catherine about the young man who was making such a reputation back in Basle; and, though there was officially no room for another mathematician in the academy which Peter had set up, Euler was smuggled in as a doctor. Soon afterwards he married and settled down to a programme which included the supervision of Russian weights and measures and elementary mathematical education as well as an enormous amount of creative work.

This lasted until 1740, when Euler was invited to do a similar job in Berlin for Frederick the Great – another enlightened ruler and one who, like Peter, had been advised by Leibniz. The mathematician's 'useful' work there included the planning of anything from a national canal network to pensions schemes: today's examples of these latter, such as are embodied in Britain's welfare-state system, largely follow Euler's calculations as to the payments needed to provide for future demands. But Russia still had claims on him, one unexpectedly pleasant outcome of his triple nationality being that, when Euler's borderland farm was damaged in the 1760 frontier dispute between Russia and neighbouring Prussia, each side outbid the other to recompense the great mathematician.

In fact, Russia wanted Euler back in St Petersburg; and when six years later another Catherine – the Great – was ruling there, he returned on an invitation which was backed by the offer of a house and eighteen servants. A few years later, when his private library was accidentally burned out, it was regarded as a Russian national duty to see that every paper was in some way replaced.

In 1783 Euler met his marvellous end. If it is asked 'but what did he *do*?' one answer is that he wrote over 600 'papers' – some of them still unpublished, most of them longer than a chapter of this book, and all of them dense with creative thinking. In these Euler added to virtually the whole of the mathematics of his time and much of its physics too. From this mass of work we can take only three examples.

In almost any book on the nature of mathematics one will meet diagrams like the pair here. The first shows how seven

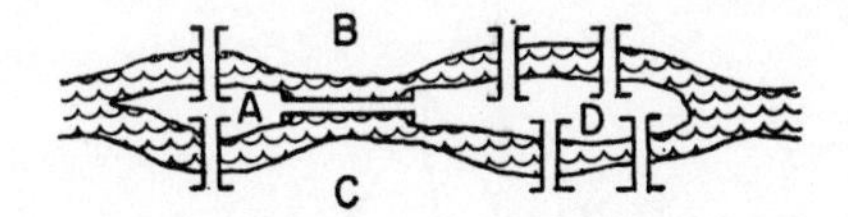

bridges linked the little two-island city of Königsberg where Regiomontanus had worked out his trigonometrical tables – a city which in the eighteenth century was in eastern Germany. There was a standing challenge there to devise a walk which took in a crossing of every bridge *only once*. And though the inhabitants were coming to believe that this was impossible, nobody could explain why this was so until the problem was referred to Euler.

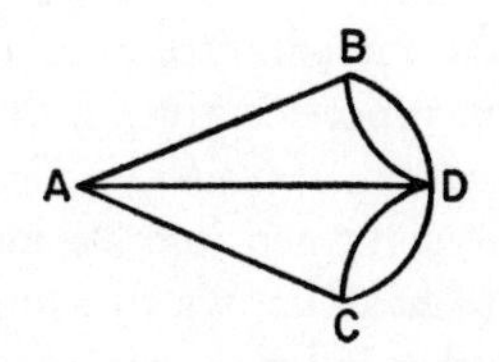

The mathematician's first step was to produce the second figure, in which the bridges are represented by distorted lines or 'arcs' and the shores and islands – which we are not much interested in – by meeting-points, nodes or 'vertices'. Now (reasoned Euler) a figure which can be traced without pen leaving paper is the equivalent of a continuous walk and can have any number of *even* vertices, since these correspond simply to a track crossing an earlier one. But it can have at the most two *odd* ones, equivalent to a starting- and a finishing-point. And since the Königsberg network has three vertices with three arcs and one with five, it is not 'unicursive'.

This may seem as trivial a business as Pascal's dice-throwing problems. In fact, it turned out to be just as important as the starting point for a new kind of mathematics. Thus, Euler

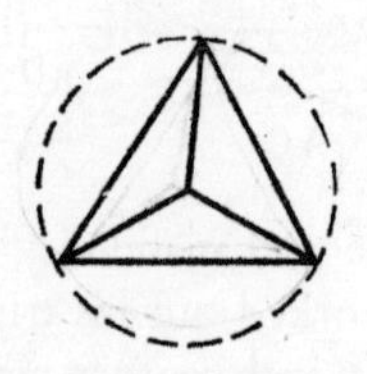

soon extended his rules to cover networks in general, and demonstrated how the edges and vertices of a solid figure could be distorted into such a plane network. (The figure here, for instance, represents a flattened-out tetrahedron.) Then, by a simple but ingenious argument, he showed how all such networks could be reduced to a triangle by getting rid of edges, vertices and faces in a limited number of ways.

A summary of this process is given in 'Euler's rule', which states that, in any assembly of planes, vertices and edges which can be drawn on or deformed into a sphere, $P+V = E+2$. Archimedes seems to have known of a similar formula; but what is important is that Euler's approach gives us a new way of looking at solid bodies. There is, for instance, a figure called a 'torus' or anchor-ring, which is shaped like a doughnut or quoit; and this does *not* obey Euler's rule but instead one of its own. The later Greeks had been fascinated by the torus: here now was proof that it was fundamentally different from a sphere in a way in which a cube was not.

We have seen how the projective geometry of Desargues had moved away from Euclid towards being *non-metrical*: it was interested only in whatever remained constant when (say) a triangle was squashed into a different triangle. This new geometry went a stage further, and concentrated on what remained even when the triangle was transformed into a circle – in the most fundamental qualities such as order and continuity, division and separation. It was the geometry of rubber sheets, balloons and elastic bands – and also of knots, maps, mazes and the like. The subject was at first called *analysis situs*: later, as we shall see, it acquired a handier name and a greater importance than even Euler could have predicted for it.

So this great Swiss mathematician, virtually single-handed, created a new geometry. The second aspect of his work which we must consider here takes us back to the common ground of the eighteenth century and was a *de*-geometricisation.

Readers may have been struck by an apparent contradiction earlier in this book, in that we described as advances both the Arabs' setting free of algebra from the chains of Greek geometry and Descartes' reunion of these two domains when he drew out his equations as curves or graphs. But we *did* refer to the Cartesian framework as a 'scaffolding'; and even before Euler's time, a century later, pure mathematicians had begun to strip away such visual aids.

It was only those (like Newton) whose interests were primarily mechanical who envisaged – say – the parabolas of motion as physical trajectories like smoke trails left behind a rocket. To others they were assemblages of symbols to be handled according to certain rules; and much of Euler's work was devoted to developing these rules. Just as the Arabs had first liberated algebra from figure-making, so this new abstraction had the advantage that un-drawable equations involving four variables could be handled as confidently as those involving x, y and (at a pinch) z which *could* be drawn.

Here Euler helped algebra to stand on its own feet again, and taller than before. One of his greatest talents was that skill as a manipulator of symbols which alone would justify the christening of the fundamental constant 2·718 . . ., etc., as Euler's number or *e*. In particular, he had an uncanny ability to spot the short cuts or algorisms which enabled an apparently new problem to be handled by the substitution of equivalent formulae, perhaps drawn from some other branch of mathematics such as trigonometry. And this helped Euler in another achievement which involved much thought even if it brought comparatively little glory – the tidying-up of the entire infinitesimal calculus.

The concepts upon which this rested, though roughly according with commonsense and certainly no vaguer than the familiar Euclidean definitions of points and lines, had never appealed to the mathematical logicians who smelt sleight-of-hand in the 'now-you-see-it, now-you-don't' trick which is at the heart of the traditional presentation of infinitesimals; and so since before 1700 there had been a move to rest the calculus on a reasoning which was less easily challenged if (unfortunately) harder to follow. This movement, today complete, owes a great deal to Euler, who also tightened up the links between its integral and differential aspects and showed the importance of the idea of *continuity* in the calculus of Newton and Leibniz.

Such criticisms, however, worried only the purists: practical men were prepared to take the infinitesimal calculus on trust for the sake of its enormous and reliable usefulness. And that usefulness, as even Oresme had realised, was not confined to problems in which there was a 'flux' of physical conditions but embraced those where there was simply a spectrum of *possibilities*.

For instance, consider the difficulties of a man who has a sheet of metal one foot square and who wants to make from it a tank with the largest possible capacity. His first thought is to waste as little material as possible by snipping away the

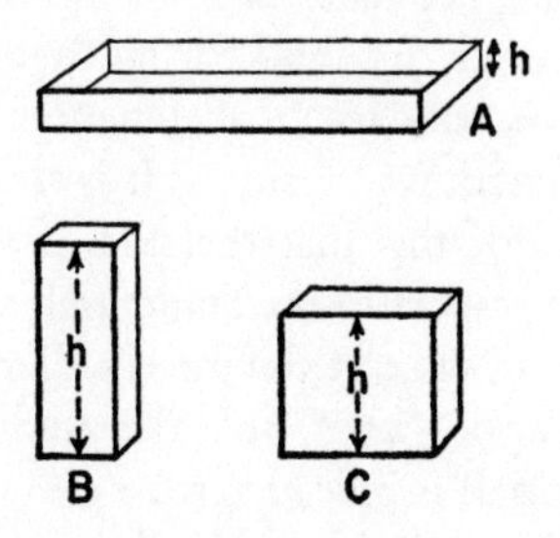

corners and soldering the rest up into the shallow tray A. Just as he is about to snip, though, a kindly friend points out that such a trough is not far removed from a flat surface which would contain nothing at all, and suggests that he might do better with the small-based but deep tank B.

To decide who is right the two men could draw a graph showing how the volume of a tank (V) is related to the length of the sides of the little squares (h) which must be cut from a larger square of metal in order to make it: this graph would in fact take the form of a parabola with its peak corresponding to a division of each side into thirds, suggesting a soldering-up into a cube (C) for the best result. But using calculus methods there is no need for any such sketching. We simply start with an equation (which the reader can produce for himself, assuming the sheet to be one unit square) expressing V as a function of h. We differentiate this, and then equate the 'derivative' to zero because (as Fermat first realised, and which is the key to the method) *when a Cartesian curve reaches a maximum or minimum value, its gradient is nil.* Finally we solve the equation. The infinitesimal calculus is indeed the perfect instrument for handling all such problems, even though they do not themselves involve any physical change.

This is a simple example of the 'calculus of variations', an idea which in essence dates back to the Greeks. Similar to it is the 'calculus of errors', and for an example of what *this* is about we can look back to our sketch on p. 100 of the graph of the 'periodic' function y = sine x. It is clear from studying a complete 'cycle' of this that the sine of an angle changes most rapidly with respect to the angle itself when the latter is about 0°, 180°, 360°, etc.: at the intermediate maximum and minimum points at 90°, 270°, etc., the gradient (corresponding to the differential) is indeed zero.

Now, in experiments with physical apparatus scientists are often measuring *indirectly* – estimating an angle from its observed sine, for example, or alternatively having to measure an angle when it is the sine which really interests them. As the example above shows, we should be much more accurate in gauging an angle from its sine if it were close to 0° rather than 90°; but the latter extreme would be a better area to work in if we were measuring angles with a view to using their sines. The calculus of errors is hence a tool which enables the working scientist to get the best from his instruments.

Other calculus developments included methods of multiple integration, and of the 'partial' differentiation used when one variable had to be sorted out from several interdependent ones. But although Euler is particularly associated with the calculus of variations we have now entered the main field of eighteenth-century mathematics; and this is itself a field where many men carried out investigations so closely-linked that, unless we first consider their work as a whole, we are likely to feel like a schoolmaster in charge of a class with everybody shouting at once. Several of these men (as we have seen) were Swiss: a number were Italian: a few were German: and even Japan contributed ideas. In fact, the only leading mathematical nation of the time barely represented was, as a result of Newton's vast but now unprogressive influence, England.

One of the first of this group of workers, for instance, was Abraham de Moivre, a religious refugee who escaped from Paris to London. There he was at first helped by Newton himself; but later he was reduced to scraping a living by solving mathematical brain-teasers in coffee houses, so few were those who could understand his more serious ideas. In despair, de Moivre simply went to sleep until he died.

The greater part of this new work was carried out in France, though; and now we must try to give an idea of its scope. In an earlier chapter we saw how the seventeenth-century mathematicians found numerous (and sometimes surprising) links between trigonometrical ratios, logarithms, and the constants such as e and π which emerged from the sums of series: functions generated by calculus methods now appeared to be woven into the same pattern, and it was also one in which i frequently appeared. The eighteenth century later added to this list a number of other expressions, such as the 'elliptical' and 'hyperbolic' functions which in some respects bore the same type of relationship to these figures as ordinary sines and the like did to the circle: originally devised purely as mathematical exercises, such functions varied over two 'periods' instead of one and so came to be used in the study of certain complex relationships.

The methods used for the manipulation of all these symbols were essentially the same as those used in – for example – the solution of ordinary equations. This aspect of algebra did, however, focus attention on some special problems such as the nature of series. Why does the $1/n^2$ series, $1+\frac{1}{4}+\frac{1}{9}+\frac{1}{16}\ldots$ converge to a fixed limit, for instance, while its $1/n$ relation, $1+\frac{1}{2}+\frac{1}{3}+\frac{1}{4}\ldots$ never reaches one? Why do similar-looking series sum-up to integers, to roots or to numbers like π? And how can we distinguish between the various types? Much of the mathematics of the next two centuries was to be devoted to such thorny questions, some of which are still not completely solved.

Since at the same time 'traditional' algebra was also breaking great tracts of new ground, it is not surprising that a split came between the two branches: this division seems to date from the work of the fine but short-lived Scots mathematician Gregory. What is regrettable is the inappropriate and confusing names which they were given. For the term 'algebra' as used in higher mathematics became confined to the handling of equations and to similar processes involving only rational powers, coefficients and constants. It was indeed so strictly applied that numbers such as π and e became called 'non-algebraic', even though their investigation gave rise to books full of the symbolic manipulation everywhere recognised as 'algebra'.

If a problem involved infinite series, generalised sines and logarithms, and the like, it was now said to belong to the realm of 'analysis'. This term (first proposed by Vieta as an alternative to 'algebra' in general) is completely misleading; for it means simply a 'breaking-down' and the only proper use of the word in mathematics is in the sense in which it has hitherto been employed in this book, to imply the extraction from a physical problem of its symbolic content. There is certainly no justification except the usage of word-blind mathematicians for saying that the problem of Lombardo's loan belongs to analysis and that that of Ali's sheep to algebra. But since the usage *is* established we shall have to employ it occasionally from now on, and save our integrity by putting 'analysis' in inverted commas wherever it is used in the rough sense of 'serial algebra'.

If the pages of this book had to keep in step with the growth of mathematics, then the sheer bulk of eighteenth-century 'analysis' would demand that it should be given at least a chapter. Quite apart from its difficulty, however, this is very much a 'mathematicians' mathematics' whose meaning – and whose relevance to anything outside its own world – is not at all easy to translate into words. To give the flavour of it, let us quote only one example of 'pure analysis' before seeking how such results were put to work.

This example concerns the relation between e and π, those fundamental constants whose symbolism (like that of i, and of much more of the modern language of mathematics) had been standardised by the great Euler. As early as 1714 an Italian mathematician derived the formula $\pi = \dfrac{e}{\sqrt{\log_e 1 \times \log_e i}}$, but even this was not as surprising as that $e^{i\pi} = -1$. One of the men who hit on this latter relationship between the four most important numbers in all mathematics said that he did not understand it and barely believed it – but that he knew that it must be true since it was backed by unchallengeable logic.

For what does this statement mean when it is – very crudely – interpreted in words? We start with a number which can never be accurately written-out. We take another number of the same type – first met with as something to do with circles – and subject it to a process which can at best be visualised as a kind of twisting in space. We then take as many of the first number as are represented by the second one (after 'twisting'!) and multiply them up. And as a result of all this, we are told, we reach minus one – an expression which was itself once regarded as all but meaningless.

The almost mystical overtones of such formulae appealed to some men in an age which still liked to look for theological clues in its symbols. But not all of 'analysis' breathed such a rarefied air, and the results of much of it – its advances in solving equations involving differentials and integrals, for instance – were put to work in that perfection of the laws of energy which was one of the main achievements of the second half of the eighteenth century. The greatest name here is that of P. S. Laplace, a man who did not care where his mathematics came from if they helped him towards interesting results.

We have seen that continental Europe was late in accepting Newton's vision of the way in which his gravitational law ruled (or, at least, described) the behaviour of the universe: France, in particular, clung for nearly a century to a peculiar system of 'vortices', worked out by Descartes, which purported to explain the movements of atoms and planets. By 1775, however, Laplace had accepted Newton's basic ideas and had rethought them with the help of a calculus notation more flexible than that which the Englishman had devised.

As a man this son of a Breton peasant was not at all admirable, becoming renowned for the place-seeking and political intriguing by which he clawed his way to honours. But his life's work, the *Celestial Mechanics* published between 1779 and the eve of his death in 1827, was a book whose influence in its own field was second only to that of Newton's *Principia* – for it virtually founded modern cosmology, the study of the evolution and nature of the universe. Here even the limited solar system presented some problems which turned out to be as difficult as they appeared simple: for instance, eighteenth-century workers long struggled without success to resolve the way in which a mere three bodies at large in space (such as the sun, moon and earth) would affect each other's positions and movements through gravitation. But Laplace made progress not only in this direction but with matters to which Newton had barely turned his great mind, such as a consideration (vital to any investigation of the history of the universe) of whether disc-shaped and pear-shaped balls of matter, spinning round, would hold together or break into two or more parts.

He also tackled the question of whether the sun's system as a whole was stable or in process of change. But although astronomy was to remain an important science it was by the eighteenth century no longer *all*-important, and there were problems enough on earth to be solved with the aid of man's new mathematical tools. The Bernoullis, for instance, had applied these to questions concerning the flow of fluids, and so had founded the very important (if mathematically complex and rather ugly) technologies of hydrodynamics, aerodynamics and rheonics, and a little later – and associated with the name of the Joseph Fourier who also made important advances in the theory of heat – came a rather similar analysis of wave-motion.

This was, perhaps, even more important: it was certainly to be of far greater influence than the men of the eighteenth century could have guessed. Even then, however, it was realised that in addition to visible waves like those on the sea there were rhythmical changes which could be drawn (or even, with the aid of simple mechanical devices, draw *themselves*) as repeating shapes which represented musical notes, the rise and fall of the tides, and so on. And, following Huygens, many suspected that light and heat might in some way be associated with wave-motion too.

The graph of a simple sine function – A in the diagram – represents the purest form of wave-motion. But in nature we far more often meet cases where several sine- waves have been mixed together. For instance, the wave B (which differs from A in its height or *amplitude*, its *length* or period of repetition, and its zero-point or *phase*) can be superimposed on A to give the composite shape C.

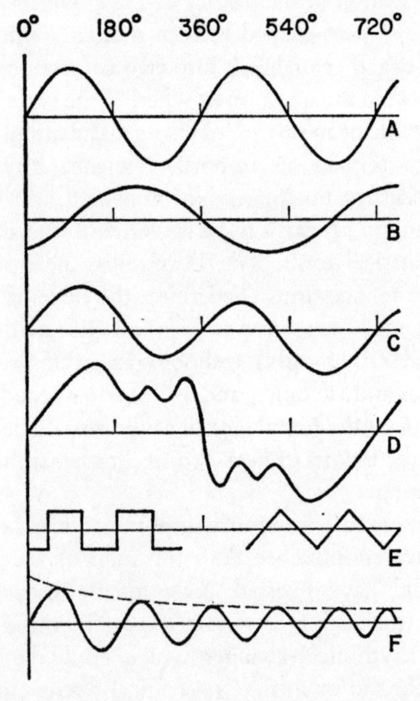

Very early in this book we saw how musical instruments generate harmonics or overtones; and though flutes and violins produce fairly pure sine waves, the richer sound of an oboe or 'cello demands to be represented by a complex curve like

D. Similarly, the rise and fall of the tides approximates to a sine form in mid-Atlantic but not in a tortuous estuary.

What Fourier showed was that *any* self-repeating or periodic curve could be represented by a series of the form: A_1 sine $(B_1\ x+C_1)+A_2$ sine $(B_2x+C_2)+A_2$ sine (B_3x+C_3) ... where A_1, A_2, etc, represent the relative amplitudes of its components, B_1, B_2, etc., their wavelengths, and C_1, C_2, etc., their phases. Even angular curves like the two examples at E can be described by such a series – though in these cases an infinite number of terms is needed. By a modification the formula can represent waves which gently die away as at *E*.

The 'Fourier analysis' of wave motion – which even a century ago was being aided by machines – is one of science's most versatile weapons: it is used, for instance, in unravelling the very complex cycles of natural changes which affect our climate. But it would be unfair to leave such themes without recognising the debt which men like Laplace and Fourier owe to their predominantly 'pure' colleagues – and especially to three fellow Frenchmen.

Fourier himself had been an orphan educated at a church school where he 'ghosted' highly intellectual sermons at the age of twelve, ran the streets and generally behaved like a young delinquent at thirteen, and a year later still was engaged in secretly reading mathematics through the night. But at least he was never *thrown* away as a child, as was the new-born baby discovered one November night in 1717 by a gendarme patrolling the old Cité area of Paris when he turned his lantern on the steps of the chapel of St John the Baptist 'le Rond'.

The boy was christened after the church, and poor but devoted foster-parents were found for him. Later he added, on his own account and for some secret reason, the surname of d'Alembert. And under that name he became famous not only as a mathematician but as one of the *savants* who compiled that great distillation of eighteenth-century knowledge of (and outlook on) the world known as the French *Encyclopaedia*.

Nineteen years later Joseph Lagrange entered the world in happier circumstances. Another of mathematics' minor nobility, the shy, modest and generous Lagrange had influential connections all over Europe (especially in Italy, from which his family came) and was also a distant relation of Descartes'. At the age of sixteen he had already became a senior lecturer, and when his friend Euler left Berlin to return to Russia, Lagrange took his place at the court of Frederick the Great. The emperor then said that the greatest monarch in Europe needed its greatest mathematician: this judgement was of course unfair to Euler, for if genius can be measured the two men were at least equal.

Lagrange remained in Berlin for twenty years, where in addition to producing pure mathematics concerned with such subjects as the calculus of variations he worked on devising the first real advance in its field since Babylonian times – the metric system which carried the 'radix-10' principle from computation into weights, measures and currency, which conquered most of the world (and all of the world of science) in the following hundred years, but which is only now being generally adopted in Britain. Sensible as this system was, though (and it was one of the few achievements of royalist France to survive the coming horrors), its inventors made the mistake of believing that its basic unit of length should be linked to the diameter of the earth; and hence Lagrange, like several other French *savants* of the time, wasted years in surveys in uncomfortable climates.

He nevertheless lived until well into the last century – though dying before the concept of the metre as an arc of the earth's surface gave way to its more sensible definition as the length of an arbitrary metal bar. In Lagrange's later years the gentle melancholy which had ruled his life (one of the few who could lift that gloom being Marie Antionette) gained the upper hand, and like Newton he lost almost all interest in mathematics. Among the miscellaneous achievements of this much-loved man, whose last thought was to reassure his friends that dying did not hurt, were advances in the theory of sound and the recognition that, in Newtonian mechanics, time had become just another 'dimension' like the three of ordinary space.

The next in a line of men of great distinction if confusingly similar names was to be A. M. Legendre, an all-round mathematician rather than a pure 'analyst'. Legendre, however, belongs to the very end of the period covered in this chapter; and in any case it would be false to give the impression that this school of mathematics had it all its own way.

For example, France (again) produced a great geometer in Gaspard Monge. Of a practical turn of mind – for at the age of fourteen he had built a useful fire-engine for his native town of Beaune, and two years later had accurately surveyed every inch of its streets with home-made instruments – Monge developed the ideas of Desargues and became the inventor of the modern system of engineering drawing: this he applied to designing those star-shaped forts which were typical of France around 1800, with the result that his friend Napoleon classified 'descriptive geometry' as a state secret. To take a leap forward in time, Monge was also the inspirer of Poncelet, a military engineer who nearly died in Napoleon's retreat from Moscow and who kept his mind active during his convalescence by re-creating mathematics with no other tools than the walls of a prisoner-of-war cell and sticks of charcoal from his stove. Poncelet is especially remembered for his work in developing projective geometry to a stage where it almost absorbed the former geometry of angles and distances.

The algebra of the eighteenth century (to which Lagrange was a major contributor) was still very largely what it had been a thousand years before – the study of general methods of solving equations, either for their own interest or in the service of science. When speaking of the Arabs, for instance, we introduced the idea of 'simultaneous' equations, a system of any number of these being soluble if there are no more unknowns than equations. And taking a pair of the simple 'linear' form and expressing them as $A_1x+B_1y = C_1$ and $A_2x+B_2y = C_2$, we can easily work out general solutions for x and y in terms of the two sets of coefficients and constants.

Now, it turns out that these solutions (which were first reached in China and Japan) are bespattered with formulae such as $A_1B_2-B_1A_2$. This particular expression can be written in the form shown here, when it is known as a *determinant*. The notation at first appears a pointlessly clumsy one; but as one

$$\begin{vmatrix} A_1 & B_1 \\ A_2 & B_2 \end{vmatrix}$$

passes on to systems of equation of the quadratic, cubic and higher orders it becomes more and more useful. From it there developed the generalised 'matrix' algebra by which groups of coefficients could be handled like individual ones, and also techniques for deriving from equations or groups of equations their *invariant* qualities – i.e., those which we visualised in our discussion of Cartesian geometry as the properties of a curve unaffected by shifting its axes in various ways. As we then mentioned, this type of algebra was to prove immensely useful to physicists working a century or more later.

Finally, from the arithmetic of the period we can select a theme, itself dating back to about 1650, whose applications are easier to appreciate. When at the start of this book we looked at the basic properties of numbers we saw how the series of squares could be built up by addition. Conversely, of course, it can be broken down by repeated subtractions as is shown, until after two stages a constant difference of 2 is reached between successive terms.

0		
	1	
1		2
	3	
4		2
	5	
9		2
	7	
16		

What happens if we apply the same treatment to the series of cubes? Again we reach a constant, but this time it is equal to 6 and only comes up after *three* subtractions. And after one or two more trials we shall spot the general rule that any such 'geometrical series' ($y=x^n$, for positive integers) can be 'differenced' in n stages to a constant value of 'factorial n' or $n!$ Thus, for x^4 the final fourth-stage difference will be $4 \times 3 \times 2 \times 1 = 24$.

The theoretical interest of this process is that it provides an analogy to the process of differentiation (it is, in fact, an example of the 'calculus of finite differences'), and that all such links between the mathematics of the continuous and that of the discrete are important in understanding the nature of numbers and operations. Its practical value lies in that such calculations provide a convenient method for producing tables useful in astronomy and similar arts.

Supposing, for instance, that we are compiling a list of

fifth powers. By the time we have reached $5^5 = 3125$ we will be getting tired of multiplications; but by drawing up a 6-tier pyramid of differences and working back from the final 120 ($= 5!$) by adding this to the last number in the previous line, and so climbing the ladder, we will after five additions arrive at 7776, which is 6^5. Useless for calculating *individual* values of a function, this method (which it is well worth while the reader trying out) is ideal for computing lists of them. And it is particularly well adapted to the use of adding machines.

The above account gives a few glimpses into the mathematics of the eighteenth and early nineteenth centuries. If, however, we have covered this period with only a single reference to its overwhelming political event, the French revolution, this is not because the mathematicians of what was then the nation foremost in all branches of our subject were unaffected by the Terror. For though none was guillotined, one was poisoned in mysterious circumstances and another had a very close escape from the block and blade.

This lucky survivor was Monge, who had earlier done much to provide the revolutionaries with their arms. With the establishment of Napoleon's dictatorship Monge came into his own, his reckless courage alone setting him at the head of that circle of scientists with whom the 'republican' emperor surrounded himself – and whom he honoured with titles and appointments. (Monge, for instance, was asked to found the *Ecole Polytechnique* which dominated French science for many decades.) But this militaristic period was also the last in which the French superiority in mathematics went unchallenged.

Shakespeare had written of Caesar that he did 'Bestride the narrow world, Like a colossus'. The phrase has been applied to other conquerors, from Alexander to Napoleon himself; but in fact no military man has ever deserved the compliment so well as have a handful of pioneers pre-eminent in the history of ideas. And of that handful, none merits it more than a quiet German who was first making a name for himself when Napoleon declared war on the world, and who was to be acknowledged as a mathematical genius without equal by the time his senior by eight years slunk away, a beaten man, leaving his troops to rot amid the blizzards of Moscow.

6

The Great All-Rounder

At the close of the eighteenth century, and within fifteen years of each other, two men appeared almost out of nowhere who were destined to be numbered among the giants of science. In a companion book we met Michael Faraday and saw how he emerged from an unpromising background; but at least his London was a centre for research into physics. By contrast, the German town of Brunswick seemed in 1777 little better equipped to produce the greatest all-round mathematician of all time than did the family that from which came that Beethoven of science, Karl Friedrich Gauss.

His father, indeed, was a casual labourer who actively resented his son's intelligence. Karl's mother was more far-sighted, though, and her brother did much to make up for the shortcomings of her husband. The boy also had the advantage – rare for that time and class – of being an only child, so that a very close link was formed between him and his mother.

We have met several 'boy geniuses' already in this book, but perhaps none was so much of a mathematical early developer as Gauss: he once said that he learned to count before he could talk, and there is a story of his correcting the household accounts before he was three. But it is from his years at a village school that the classic anecdote comes.

The master there was in the habit of setting long addition problems of the shape: $116+131+146+161\ldots$, perhaps extending to so many terms with so many digits in each that each sum could take an hour to work out. But the ten-year-old Gauss soon spotted that all these weary exercises had in common that fact that there was a constant difference (in the above case, of 15) between the terms. And so he suspected that the master had some shorthand formula which enabled him to avoid the tedious work.

What Gauss – and we – face here is a problem in summing a series. It is true that the series involved is of the simplest 'arithmetical' type and was analysed by the first of all known mathematicians, Ahmes. But *any* such problem must be cracked by real mathematical insight rather than 'brute force and ignorance'.

Generalised, the series which the teacher used looks like this: $a+(a+d)+(a+2d)+(a+3d)+\ldots(a+(n-1)d)$, where a is the first term, d the constant difference, and n the number of terms. But we can best follow the method of summation used by considering the even simpler case where a and d both equal unity. We are then trying to total-up the integers 1, 2, 3, 4 ... – with the results, of course, giving us the 'triangular' numbers.

The trick is to write down the series and then the same sequence backwards:

$$\begin{array}{llllll} 1, & 2, & 3, & \ldots & (n-2), & (n-1), n \\ n, & (n-1), & (n-2) & \ldots & 3, & 2, \quad 1 \end{array}$$

Each vertical pair of terms then adds up to the same total of $n+1$; and since there are n pairs, the total of the whole double-series must be $n\ (n+1)$ and the triangular numbers themselves be given by a half of this formula. It is all beautifully simple once one has seen it done, and the extension to Gauss's schoolroom problem follows with little more fuss. But a highly-intelligent man might think about the difficulties of summing such an *indefinite* number of terms for days – and then give up.

Young Gauss, however, handed in a slate with the right

answer on it almost as soon as he met this type of problem. Far from being annoyed that his pupil had short-circuited the task he had been set, the master recognised that he had a genius on his hands. No mathematician himself, he bought Gauss the best textbook he could find in Brunswick – which the boy mastered in a few days – and then passed him over to a seventeen-year-old assistant who was also fascinated by mathematics (and was to win real distinction in it). For the next few years the two worked together like Descartes and Mersenne; but soon the younger was well in the lead.

At fifteen Gauss was taken up by the Duke of Brunswick, who found him a place at a more advanced college. There he developed that gift for languages which often accompanies a talent for handling symbols: both, of course, are aided by a good memory, and Gauss himself could remember the million-odd digits of an ordinary logarithm table without conscious effort. He also had a lifelong interest in world literature, if a rather 'middlebrow' one: his favourite author in English was Sir Walter Scott, but he could not accept Shakespeare's unhappy endings.

In these high-school years Gauss also interested himself in some of the 'observational' sciences. But at the same time he was making really important advances in all three of the main divisions of mathematics. In geometry, for instance, he had already formed the basic ideas whose development we shall be looking at later. In algebra and 'analysis' he had spotted fallacies which had gone undetected for a long century and had gone back to the roots of the subject to found the cult of 'rigour' which – again – will call for later attention. And in pure arithmetic, his favourite subject, he showed how a new notation could help solve problems which had baffled Diophantus, Fermat and their successors.

Because this notation is introduced nowadays in some

school courses – and also because pure arithmetic must have rather a poor showing elsewhere in this book – it is worth giving a paragraph or two to this development. One way of approaching it is to note that, in the series of the squares of the integers (1, 4, 9, 16, etc.), only the digits 0, 1, 4, 5, 6, and 9 ever appear in the final place: no squares end with a 2, 3, 7 or 8. Now, since we are using a denary notation these final digits represent the remainders or 'residues' when the squares are divided by 10, such a divisor being in this kind of arithmetic called a 'modulus' or 'modulo'. And Gauss realised that to investigate such problems as why some numbers were 'quadratic residues' and others not, he needed a notation which drew attention to the fact that (for instance) both 12 and 27 leave a residue of 2 when divided by 5 – and which completely ignored what ordinary practical arithmetic is most interested in, the number of times which 5 'went into' the numbers.

A typical statement in this novel type of arithmetic hence reads: $16 \equiv 2 \bmod 7$. Such 'congruences' (a term unhappily stolen from geometry) lead to 'finite' arithmetics which involve sets of only as many numbers as the modulus itself demands, and which have their own addition and multiplication tables. The subject is not at all remote from everyday thinking, for the idea of a modulus is implied whenever – for instance – we add a day of 24 hours to 5 o'clock and get, not 29 o'clock, but 5 o'clock again. But its importance is that (with reservations) mathematicians can handle these congruences like ordinary algebraic equations and so prove some simple-looking but very thorny theorems involving divisibility. By using them, for instance, Gauss was able to give six independent proofs of a problem which had defeated even Euler – and, still more important, help to weld the theory of numbers into a consistent whole for the first time.

The amazing thing is that he did all this whilst he was still at high school and undecided as to whether or not to make his career in mathematics as opposed to the study of language. But around his twentieth birthday Gauss made two discoveries which swayed the balance. One was a proof that every integer is the sum of three triangular numbers (or, as Gauss joyfully expressed it in his note-form diary, EUREKA! num = $\triangle + \triangle + \triangle$. The other strengthened a bridge between Euclidean geometry and pure arithmetic formed by the Greeks by showing how the problems of constructing regular figures with a ruler and a pair of compasses were linked to the number of sides in the figures. Neither theorem was of the slightest use outside its own world: both stand high among the achievements of the human mind.

So, with a reputation already made, Gauss entered a university in 1795. The university was Göttingen, one of that handful of old, quiet and pretty small towns near the Rhine whose academic establishments were little more than that of a single college at (say) Cambridge. But – almost solely as a result of the inspiration which Gauss brought to it from his freshman year onwards – Göttingen became one of the world's great centres of mathematical inquiry, and held that position until Hitler's expulsion of the Jews in the 1930s ended Germany's claim to be a leading mathematical nation.

Although Gauss was supposedly a student, his three undergraduate years were spent almost entirely in carrying out original work – and, in particular, in writing a classic book on that mathematics of integers and rational fractions which is considerably tougher than the parallel mathematics which admits *any* numbers. But his printed work never represented more than a tiny part of his thinking, for Gauss refused to make a discovery public until he had worked it over and over and reduced it to its purest form. He was proud of this practice, and took for his unofficial coat-of-arms a tree bearing a handful of fruit with the motto 'few, but ripe'; but it is arguable that it was in fact a rather selfish habit which delayed the growth of mathematics and meant that men of later generations had to struggle with problems which Gauss had solved even before he came of age. Another of his traits, however, helped the spread of knowledge: for long after his contemporaries had given up Latin in favour of that babel of local tongues which is so ironically characteristic of our 'international' age, Gauss still wrote his publications in the ancient language of science.

Another irony was that the very success of his first book – the *Arithmetical Disquisitions*, published in 1801 when its author was twenty-four – meant that Gauss could no longer devote himself to pure mathematics. The generous Duke of Brunswick (himself soon to die after despicable treatment by Napoleon) had offered him a pension for life, so that Gauss could, had he wished, have spent his time remote from practical problems. But by temperament he wanted to be financially independent, and he was also genuinely interested in that world of science and even engineering from which problems were beginning to be passed to him. For more than fifty years – for the whole, that is, of the first half of the last century – Gauss was thus posted at the centre of the world of physics, with every problem which had perplexed others in half-a-dozen branches of it being referred to his analytical mind and to his instinct for striking at the heart of a problem.

The earliest such problem was almost as trivial as it was tedious, for the calculation of the orbits of the first of the newly-discovered minor planets could have been carried out (if less economically) by any competent 'computer' who had read Laplace – the Frenchman who was among the first to recognise Gauss's greatness, and who indignantly paid a fine which Napoleon had inflicted on him for his crime in being a German. Astronomy was still a fashionable subject, though, and from his initial success Gauss won an increased reputation outside the world of pure mathematics. The tragedy was that he did not then bow out, but instead let himself be distracted by such astronomical work for two more decades.

In 1807 Gauss took up a university appointment at Göttingen; and in the next year, when he was thirty-one, he married. But tragically soon his beloved wife died, and though he soon remarried his family life afterwards was never altogether happy. In this period he produced five children and at least one great theorem in pure mathematics, that which showed that a host of 'analytical' functions, including those which gave rise to trigonometrical ratios and logarithms, were only special cases of the same series. But he did not do anything significant on the applied side until, after 1820, he descended from the heavens to an earth which was to prove much more fruitful for him.

The early years of the nineteenth century had seen advances in geodesy, or scientific surveying, comparable to those which had taken place in the mapping of the oceans after 1500. Quite apart from the problem of accurately measuring the earth's diameter which had been highlighted by the adoption of the metric system, for instance, we have seen how Napoleon's France discovered military implications in geometry. Threatened by invasion, peaceful England speeded-up a great national survey begun primarily for 'ordnance' (or artillery) purposes. And after Napoleon's defeat the states of Germany too organised a programme of improved mapping.

In 1821, Gauss was asked to advise on the surveying of north-west Germany and Denmark. The matter may sound even more of a well-paid but talent-wasting 'pot-boiler' than the study of minor planets, but there is little knowing in advance what practical problems will generate interesting mathematics. And nothing could be more typical of the two sides of Gauss's genius than that this work not only led him to invent the heliograph (a signalling instrument, using reflected sunlight, which was useful for a time in its own right and is important in the history of communications), but turned his attention towards the general properties of figures drawn on curved surfaces and the problems of 'mapping' these on to flat or differently-curved surfaces.

These themes, of course, were not new; but Gauss brought to them his characteristic approach which replaced vague intuitions with logical definitions. He was the first, for instance, to give consistent meanings to ideas such as the areas of surfaces curved either 'extrinsically' (like a cylinder which can be flattened-out) or 'intrinsically' (like a sphere which cannot), and to show how their geometries could be build up without references to any outside framework. We shall later see some of the applications of a subject which is often termed 'differential' geometry because it does for surfaces rather what the infinitesimal calculus had done for lines.

Considering the enormous range of Gauss's activities, it is convenient to note the way in which certain decades of his life were associated with characteristic themes. In 1830, for instance – when he was already over fifty years old – he moved away from geodesy and began a ten-year programme of work in the physical sciences which was to rank with the greatest of his achievements. This covered an immense span, ranging from optics to an investigation of those forces of surface tension and capillarity which make liquids apparently defy the force of gravity. But its most important aspect was Gauss's triumph in giving a mathematical structure to electricity and magnetism.

Readers of *Men and Discoveries in Electricity* may remember that the previous fifty years had witnessed the work of a line of investigators – Cavendish, Coulomb, Oersted, Ampère, Ohm and even the young Faraday – who had discovered various *ad hoc* laws about such forces. Now, a third of the way through the century which was to see electricity become man's greatest aid in power and communication, the time was ripe for these laws to be given the type of unifying treatment which Newton's theory of gravitation had given to mechanics some 150 years earlier.

And – almost miraculously – there was just such a second Newton to hand. Even in his experimental ingenuity Gauss resembled his predecessor, for together with his colleague Weber (who, like Gauss himself, was to lend his name to one of the basic units of magnetism) he designed some historic apparatus for electrical measurements; and with virtually no laboratory experience behind him the inventor of the heliograph even became side-tracked towards that fashionable technical problem of the 1830s, the improvement of telegraphy. But all this was almost trivial as compared to his work in electrical theory.

As the *Electricity* book showed, the early history of this subject was mainly concerned with stationary charges and with the fields of force – in some ways resembling, and in others differing from, gravitational fields – to which they gave rise. Then, after 1800, serious work began on the currents which represented charges in motion. Even 'electrostatics', though, produced some challenging problems: for instance, it had been discovered experimentally that if an irregular-shaped body like the one shown opposite was electrically charged, then the charge tended to concentrate where the surface was most sharply curved. And the geometry of surfaces, towards which Gauss had been led through his surveying work, was the obvious tool to give mathematical expression to this fact.

Most such work was too specialized to discuss here, but one fundamental result demonstrates how theory and practice can interact. Consider a hollow sphere which is attracting a

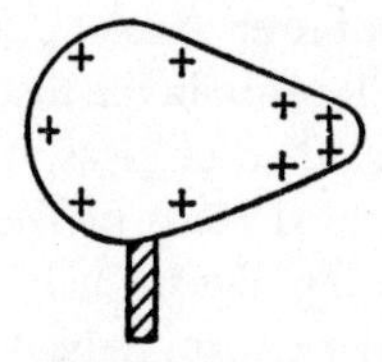

tiny body 'A' inside it through a force which acts according to some law. If this law is the 'inverse square' one which Newton discovered to be true for gravitation, then the *total* force acting on the body will be zero; for if (and *only* if) this law applies, then for any 'element' of the surface exerting an attractive force at unit distance we can envisage one x times

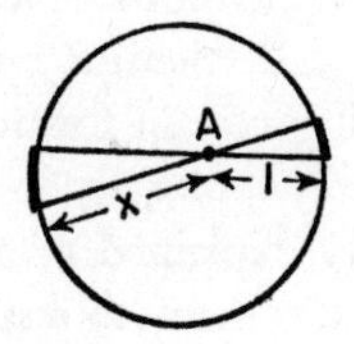

as far away in the opposite direction which will be larger in area by a factor of x^2 – but whose force will be weakened in just the same ratio.

This argument, which has roots in Newton's own work, can not only be made into a strict proof but also by means of the integral calculus degeneralised from a sphere to any hollow body. It implies that if the inverse-square law is true of electrical attraction, then there will be no field inside such a shell – and *vice versa*. But there *is* no such field (a fact which itself lies behind much modern high-voltage technology) and so the law *does* hold.

The fact that both gravitational and electro-magnetic forces are ruled by inverse-square laws, indeed, forms one of the fundamentals of our stable physical universe. (For instance, there is a parallel even in the theory of light.) Another such basic principle is that of 'least time', which Fermat developed from an idea of Hero of Alexandria's and which Gauss – himself following Euler – helped to extend into the more generalised modern form which unites much of the behaviour of radiant energy and other forces. But wherever we look in considering the physical science of the middle of the nineteenth century we seem to meet this same great figure.

It was Gauss, for instance, who realised that complex physical entities such as force and power – and electrical and magnetic units too – were built up like velocity or pressure from the fundamental 'dimensions' of length, mass and time – and that this fact provided scientists with a handy means of completing and checking certain formulae. Here, as a result of an idea which probably took only an hour or so to develop and which can be fully explained in a couple of pages, he saved future experimenters many thousands of hours of work.

Amid so many achievements it is almost refreshing to note Gauss's one failure. We have seen that static charges give rise to *fields* and steadily-moving ones to *currents*; but when charges are accelerated we meet yet other types of phenomenon, including the generation of electro-magnetic *waves* in space. In principle, Gauss realised how those differentiations from position to velocity and on to acceleration which ruled Newtonian mechanics could explain the behaviour of electricity too. But his theories were not quite general enough to lead him on to the great discovery made some twenty years later by James Clerk Maxwell – the brilliant but short-lived Scottish mathematical physicist who also applied a Gaussian analysis of nature to the physical behaviour of gases.

As the earlier book shows, Clerk Maxwell predicted the existence of radio waves some two decades before they were discovered experimentally. Today a vast body of mathematics, as well as of technology, has been erected on the foundations which he and Gauss laid. We cannot, of course, begin to look at this in any detail here; but before returning to Gauss himself there is perhaps one point worth noting.

Anybody who undertakes the study of electricity soon meets with rules such as that which states that the directions of an electric current in a wire, the magnetic field surrounding it, and its movement, are related like the thumb and first two fingers of a hand held with the fingers at right angles to each other – or like the x, y and z axes of Cartesian space. The latter analogy suggests that we should not be surprised to find $\sqrt{-1}$ turning up in the equations which represent the behaviour of electrical currents, waves and forces – and in fact this symbol is characteristic of them. But it is not there because Gauss or Clerk Maxwell 'planted' it to represent a rotation into a third dimension, but rather because at some stage of his analysis a mathematician was faced with the need to symbolise the square root of a negative quantity. '*i*' numbers might still be called 'imaginary', but they were certainly not useless now.

Gauss's work in electricity alone could be developed into a full chapter of this book; but it occupied him for little more than a decade, and even then he was continuing work in pure mathematics. After 1840 this again became his main concern, though with his seventieth birthday in sight his mind was still so wide-ranging that it could not resist any problem which came before it. Thus, quite trivial brain-teasers such as were posed by 'knight's tour' problems in chess, or by the 'magic squares' (or arrays of numbers adding up to the same total along all their lines) which had intrigued puzzlers since Babylonian times, came in for his attention at some period. And his investigations were usually so complete as to rob such problems of any serious interest for the future.

But if one were to ask which was the most influential of all Gauss's advances the answer would be found neither in his many investigations into pure mathematics nor in the obviously and directly applied work, but rather in a study to which he returned again and again during his creative lifetime of more than sixty years. Gauss himself was, as a man keenly interested in world affairs, always conscious of the potential importance of this subject, and at one time he called it 'political mathematics'. Today it is known as 'statistics' whenever that word is used in the abstract sense of a technique for handling a mass of figures rather than in a more concrete one implying the figures themselves.

Again we are in the presence of a theme worthy of its founder, for the story of the development of the statistical method demands a chapter, and almost a book, rather than a few paragraphs. As a starting-point, though, we can take the idea of an *average* or mean as presenting the simplest method of reducing an unwieldy collection of figures to a single one representative of the group. For the average to be meaningful the figures must, of course, have something in common: but it is worth noting here that they may apply to either a number of different entities (for instance, the weights of a group of boxers) or a number of different estimates of the *same* entity (for instance, the reading on a meter), and that statistics uses the same methods for dealing with both these types of situation.

For an example of an everyday problem of the first kind, imagine a man who is trying to decide whether to travel to work by train or by bus. Since the alternatives seem to rate about equal for speed, comfort and so on, he decides that punctuality must be the criterion, and so he experiments by travelling for ten days by bus and then ten by train. The bus, he discovers, is just about one minute late every day: the train was punctual on nine days, but on the tenth was *ten* minutes late.

These are quite different types of unpunctuality, and our traveller will almost certainly decide in favour of one rather than the other according to his needs. But a simple calculation of the ordinary or 'arithmetical' average of the degree of lateness gives the same result – one minute in both cases. We must hence look for a more revealing type of figure; and before Gauss came to the subject it had been realised that this was provided by a 'standard deviation' which indicated the 'spread' of a range of figures by means of another figure derived from averaging the *squares* of their differences from some mean or ideal value.

What Gauss himself did (and he again took the first, important steps while he was still at college) was to develop from this a 'method of least squares' whose implications ranged from the design of scientific experiments into electrical theory. This development implied extending the method from handling a mass of separate figures to dealing with continuous and even periodic functions – i.e. to building another of those important bridges between the worlds of the discrete and the smoothly-connected. From there it was an obvious advance to introduce the ideas of the calculus and 'analysis' into Pascal's world of tossed pennies and thrown dice; and in fact this journey had already been begun by Laplace, Legendre and others, including even de Moivre a century before.

To form a picture of what is involved let us plot the relative likelihoods of the outcomes of coin-tossing experiments on the vertical axis of a graph: the horizontal axis is divided into segments according to the types of outcome possible. Thus, the points on the top 'curve' represent the 1, 2, 1 (25 per cent, 50 per cent, 25 per cent) distribution of a two-toss experiment. The next curve down shows a similar plot for four tosses, and the others for six and eight. By now there is a clear tendency for the curve to take up a bell-like form.

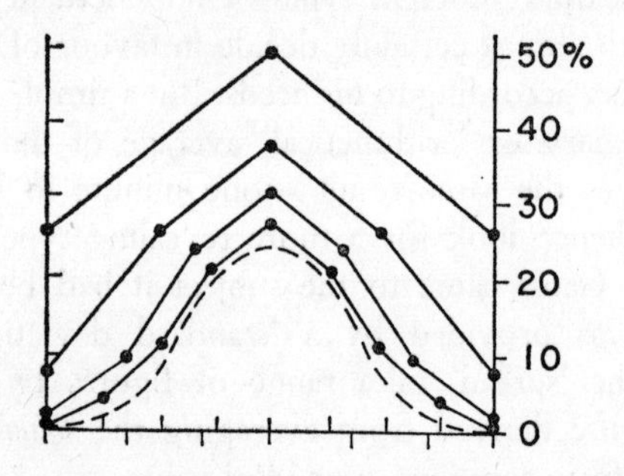

This is one way of arriving at a shape which is as characteristic of statistics as the hexagon is of organic chemistry or the parabola of ballistics: we can regard the curve as representing the limit of a binomial expansion, summarising the result of an *infinite* number of coin-tossings, or of any other process where an even chance is repeated time and again, which is approached in practice when more than a few dozen events are involved. (If *many* small ball-bearings are dribbled through an array of *many* pin-table spikes, for instance, they will pile up in this form.) Another view of the curve is that it represents an algebraic function as simple as: $y = e^{-x}$. And yet a third approach is a purely experimental one.

For whether we ask a hundred people at a fairground to guess the weight of a pig, or measure the life-spans of a thousand electric-light bulbs, we shall find that, once we have grouped our results into 'brackets' of pounds or hours and then plotted them as columns whose heights correspond to the number of results within each bracket, the crests of the columns will shape-up into such a bell-like curve. It is thus called the normal (or standard) distribution curve, and is met with whenever we are dealing with totally random deviations from some average.

If we find a curve rather similar but lop-sided or cut short, indeed, it is a sign that some special factor is involved. (Thus, we would not expect a symmetrical distribution of the deviations of train arrival times from the 'official' ones, if only because trains running early are usually deliberately delayed). This fact can itself lead to useful scientific detective-work. But it is the normal curve on whose mathematics modern statistical methods are so largely founded.

These have themselves influenced life and thought at half a dozen different points. In the middle of the nineteenth century, for instance, William Thompson (Lord Kelvin) and others realised that the energies of the molecules in a gas were scattered according to this type of distribution, so that the science of thermodynamics had to pay great attention to the statistical approach. Now, thermodynamics itself had practical applications of the first importance since it treated of the conversion of energy in engines of all kinds; and in the next century it was realised that some of its techniques were also applicable to the very deepest ideas of science, philosophy and even metaphysics. Our modern ideas of not only the nature of matter and energy but the direction of time, for instance, rest on statistical concepts.

On the practical front, statistical analyses of the type which Gauss began – and which Clerk Maxwell hailed as the ideal mathematics for the scientist or engineer – have proved of particular value in helping men to devise rational methods of 'sampling': whether these methods are applied to the performances of students in an examination or the strengths of steel bars, they can tell us how large a sample we need to take to arrive at a given degree of certainty that our conclusions will be valid ones. And this fact in turn means that there are whole disciplines of knowledge – including the 'fringe' sciences which touch on human behaviour such as economics, sociology and many aspects of medicine – which owe such certainty as they can claim to the investigator's next recourse when he is denied the possibility of *controlled* experiments, which is the use of statistics.

In addition, statistics is at the base of special techniques ranging from intelligence-testing through cipher-cracking to long-term weather prediction; and it has also extended the commercial implications of advanced mathematics from insurance into many other fields. Today it is hardly possible to launch a new pet-food without making use of this kind of research – or to manufacture the 'Doggo' efficiently without checking on its quality by (for instance) spotting deviations from some desired peak in a normal, Gaussian distribution. It is hence not surprising that a large number of businessmen as well as scientists have made themselves experts in statistics although their other mathematical equipment may be very slight.

Perhaps even Gauss did not realise the full implications of his work in founding today's statistical analysis. But this was to be one of the most famous of the achievements of a man who, though he was not without his own vanities, never sought fame. And if that sounds like an idle compliment, it is enough to say that from 1828 onwards for twenty-seven years, while the scientific world was bombarding him with invitations and honours, Karl Gauss never slept one night away from his little town of Göttingen.

Indeed, he rarely walked far outside its walls. But in 1854 he was tempted away to watch the building of the first railway line in his neighbourhood. In an incident like that which had so affected Pascal 200 years before, a coach-horse bolted and the seventy-seven-year-old mathematician received the shock which led to his death a few months later. His mind was clear, and his fight for life continued, until the last hours of a man whose total of achievements in every field of mathematics had never been approached before and may never be again.

Even outside his own group of friends, colleagues, pupils and disciples (which group included, if one exception is allowed, the first woman since classical times to make a distinguished name for herself in mathematics), the influence of Gauss was so great that for the rest of this chapter we shall be under his shadow. The difficulty is to select from a richness of material; for if the world's body of mathematical knowledge had increased some three times over in both the seventeenth and eighteenth centuries, the nineteenth was to witness perhaps a further fivefold growth before it ended. And almost every aspect of this creativeness owed something to the work of Karl Gauss.

Perhaps his most immediate influence was in developing new geometries: for instance, Gauss did much in his later years to advance the 'analysis situs' which we saw Euler creating. This work was itself carried on by August Möbius, who is especially remembered for having given his name to those bands of twisted paper, described in every book on mathematical recreations, which behave in such entertaining ways when they are slit-up. These are more than just conjuring tricks, though, for they demonstrate that our ideas about the relations between the sides and edges of a surface cannot be safely left to commonsense and intuition and that logicians were correct in questioning even Euclid's assumptions about 'betweenness'. Similarly the next name in this mathematical line-of-descent, that of Felix Klein, is associated with a '3-D' analogue of a Möbius band, a bottle whose inside becomes its outside.

The phrase 'conjuring trick' above is *literally* appropriate to some of the more amusing (if less important) aspects of 'analysis situs'. When we turn to another of the geometries which were developed from the work of Gauss we must take it in a more metaphorical sense. But this too – like so much of the 'modern' mathematics which we are now approaching – has a magical-seeming quality; and the 'Abracadabra' or 'Open sesame' to it is the word 'non-Euclidean'.

This adjective is itself a typical bit of mathematical-historical jargon which became especially fashionable in popular-science writing some two generations ago. In fact we have already met several geometries which were either *hyper*-Euclidean (like that of Descartes, who used ideas unfamiliar to the classical Greeks) or actively *anti*-Euclidean (like that of Archimedes, who was prepared to consider curves which his earlier compatriots had banned from mathematics). And even if we accept 'non-Euclidean' as the adjective best suited to a geometry of *curved* surfaces and spaces, one such had been worked out in detail by mathematicians and map-makers when they had been called on to help the mariners who had no alternative other than to sail on a curved world-surface.

'Non-Euclidean' geometry, then, was not the invention of the immediate successors of Gauss – and still less of the twentieth-century cosmologists who are so often associated with it. Its history goes back, through Mercator and the developers of renaissance 'spherics', to the later Greeks. Gauss and his followers, however, generalised from the mathematics of the sphere until they could deal with *any* surface whose curves could be expressed in symbols. And this they did partly to meet the needs of new technologies, partly (perhaps) from a sense that such results as found no use more than a century ago might do so at some later date, but mostly because the intellectual challenge was there to be met.

But why – we must now ask – should such a commonplace thing as the surface of a ball have to be treated like a strange monster? The answer is that we are so used to dealing with the *plane* surfaces to which Euclid restricted most of his attention that we often forget that these are highly-idealised (or, from another point of view, degenerate) examples of surfaces in general. Commonsense as Euclid's results appear, for instance, they would be realised accurately only on a flat earth, and they are not *precisely* valid for even so small an area as this full stop.

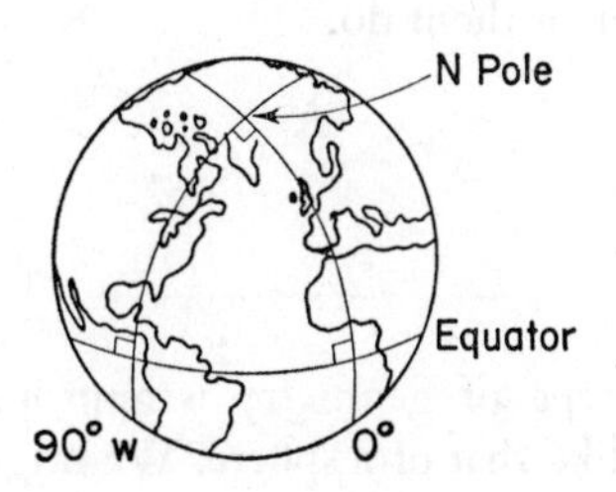

Thus, the whole concept of parallel lines which is fundamental to Euclidean geometry breaks down on a curved surface, and with it goes such theorems as that stating that the angles of a triangle add up to 180°. We hinted at this some chapters ago: now, if we take a look at a geography-class globe of the world, we can see that in the 'spherical triangle' made up of a quarter of the Equator, half the meridian of

Greenwich and half the meridian of 90°W (which passes not far from Chicago), all three angles are of 90°. In fact, the sum of the angles of a triangle drawn on a sphere can vary from only slightly over the conventional 180° for a very small or sharply-pointed figure up to almost 540° – or six right-angles – for one which 'wraps' right round the surface.

As we have seen, there was in the middle of the nineteenth century nothing revolutionary about this idea. What was more novel was the realisation of the way in which all possible surface geometries could be grouped by reference to Euclid's 'parallel postulate', which itself assumed (in effect) that through any point A one and only one line could be drawn which could be extended for ever in both directions without meeting another given line, BC. If this is so, then we have a Euclidean geometry; but we cannot prove that it is, and there remain the two possibilities that either no such line can exist or that a number of them do.

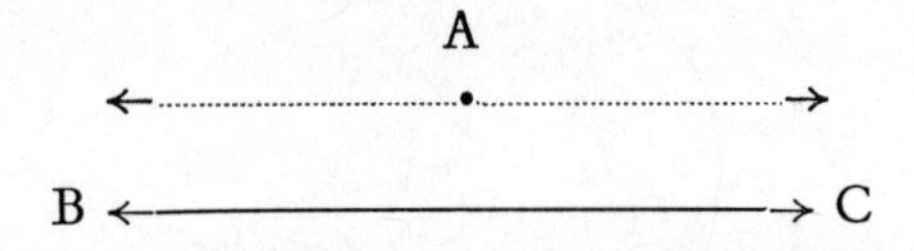

The former type of geometry is appropriate to a fully convex surface like that of a sphere. We might unthinkingly

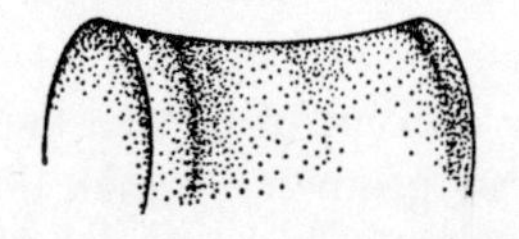

expect the latter to apply to (for example) the concave interior of such a sphere; but in fact the true converse is to be found in a saddle-shaped type of surface which is curved outward in one direction and inward in a direction at right-angles to this. This mathematics of curved surfaces was developed by the Hungarian Bolyai, by the Russian Lobachevsky (whose widowed mother had tramped hundreds of miles to find her sons university places, and who himself devoted almost his whole life to the unselfish service of a tiny university), and finally by Riemann in Germany. A sick and impoverished man but the greatest of all Gauss' pupils (had he lived longer, he might indeed have equalled his master), Riemann extended the idea of curvature into the form which is at the root of the modern idea of the universe in which we live.

So far we have considered only surfaces or 'manifolds' of two dimensions, even when these are bent round in a third dimension. But just as the Euclidean geometry of a plane can be extended to '3-D' space and thence to imaginary spaces of four or more dimensions, so we can produce a reasonable and consistent mathematics for (say) a space of three dimensions which is curved round in a fourth one.

How we are to *interpret* a space of this type is perhaps a matter for science-fiction rather than science, and there is certainly plenty to be read on the subject. But not all such multi-dimensional worlds, plane or curved, strain the imagination. The word 'dimension' is in any case a flexible one; and it is typical of the methods of modern mathematics that, having been devised to deal with situations in which 'dimension' referred first to an extension in space but was then used to cover time and mass, they were eventually applied when 'dimension' meant no more than an identification of *any* kind, a sort of laundry-mark.

Consider, for instance, the collection of all possible straight lines drawn on a plane. Each line needs four numbers to specify it (either the Cartesian co-ordinates of both ends, or those of one end plus its length and slope), and so the collection can itself be said to constitute a four-dimensional space or manifold. Similarly, the manifold of lines in '3-space' is a six-dimensional one; that of circles in 2-space demands only three dimensions; that of spheres in 3-space, four again; and so on.

Though this kind of thinking about spaces *in* spaces seems to rely on an abuse of words, it is helpful to physicists when they consider (for instance) the interactions of a cloud of molecules or electrons. Each element of these can be 'tagged' by three co-ordinates of position and by other figures representing velocities, rotations and so on, the result being an 'n-dimensional space' in which the appropriate rules can be used on the collection as a whole. And there is nothing mathematically absurd in saying that a bevy of girls, of whom one knows the heights, weights and three traditional 'vital statistics', constitutes a '5-dimensional girl-manifold'.

A geometry broadened to this degree becomes closely allied to the mathematics of *sets* to which we shall be referring in the next chapter, and indeed certain types of algebraic problem are nowadays attacked by constructing special miniature geometries. Before moving on, though, we must consider another theme linked to the great name of Gauss which has both algebraic and geometrical aspects as well as enormous importance in the physical sciences.

If (on a plane surface) we walk four miles and then three miles, how far are we from our starting point? The obvious answer is 'seven miles'; but this would only be true if we had kept going in a constant direction. Had we turned through a right-angle between the two 'legs', for instance, we should have traversed two sides of a Pythagorean triangle and so have been five miles from our start; and had we turned right round we would be only one mile away.

There are hence two quite different but equally meaningful

ways in which distances can be added. If we are interested simply in taking exercise, then we shall find the everyday kind of arithmetic (which is in this context called 'scalar' addition) appropriate to our walks. But when we get tired we may want to know how far we are from a bus-stop 'as the crow flies'; and this calls for the 'vector' arithmetic which takes account of changes of direction.

The word vector is applied to the kind of quantities, such as displacements, which are not completely specified unless we know their direction as well as their magnitude. By applying the idea of successive division by time to displacements we see that velocities are of this type, and also accelerations and

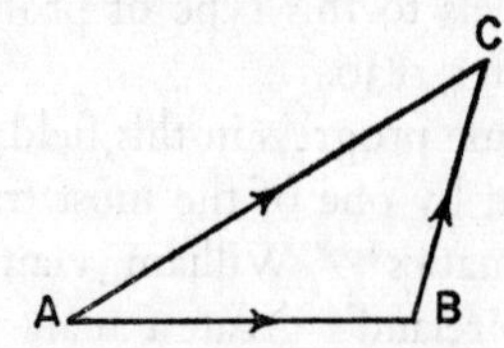

hence forces. All of these, furthermore, can be added by the kind of simple geometrical device which we have just suggested, so that if the line from A to B represent one vector, and that from B to C another, then the line AC will represent in magnitude and direction the vector sum. The scheme can be extended to 'compounding' any number of vectors, a single vector can be resolved into any convenient group of components, and subtraction is the reverse of addition so that (employing the usual notation) we could write of the above diagram $\vec{AC} - \vec{BC} = \vec{AB}$.

Such ideas as the 'triangle of forces' were established in mechanics as far back as the age of Stevinus; and though the subject is as poor in documented history as it is in rigorous proofs, it is impossible even to sail a boat without making instinctive estimates (if not formal calculations) of the way in which a course must be altered so as to off-set currents. But the forces of electricity and the like with which Gauss and his contemporaries were faced seemed to present a new problem, that of the *multiplication* of one vector by another.

There is, of course, no difficulty about multiplying a vector by a pure or scalar number: three times a velocity of 10 mph is simply a velocity of 30 mph in the same direction. To multiply 10 mph northward by 3 mph westward, though, sounds completely nonsensical. Perhaps it is, though the far-sighted de Moivre had done something of the kind more than a century before without its significance being understood; but electrical parallels to this type of problem became very practical matters after 1830.

Möbius made some progress in this field: then, a few years later, it was entered by one of the most tragic figures in the history of mathematics – William Hamilton of Dublin, usually accounted Ireland's greatest man of science though (like the Ulster-born Lord Kelvin) he was of Scots descent. Hamilton was another of those miraculously-endowed children who fill this book, for in addition to showing extraordinary powers of mental arithmetic he mastered his first foreign languages at the age of five and went on adding to them at the rate of about one a year until he was fluent in nearly fifty. He made a brilliant early career for himself as an astronomer, being appointed to a professorship before he had graduated, and when he was thirty he was knighted for his work in mathematical physics.

But there is a curse on the Irish, the curse of an over-fondness for good talk accompanied by good drink. Despite all efforts, Hamilton slipped from being a convivial man who enjoyed sharing a bottle of wine with such friends as the 'Lake poets' towards becoming the only great man of science to suffer from alcoholism. Closely linked with this problem was that of his unhappy marriage to a woman whose failure to understand him is perhaps more excusable than her incompetence as a housewife. When Hamilton's papers came to be sorted after his death, for instance, they had to be purified from the remnants of dinners half-eaten many years before.

In addition to his private torments, Hamilton suffered by coming to neglect all other mathematical ideas (such as his own important work in optics) in favour of a single scheme which he called that of 'quaternions'. One object of this was to resolve the problem of vector multiplication by constructing a space whose three dimensions were each characterised by an arbitrary 'operator' of the $\sqrt{-1}$ type. But Hamilton regarded the manipulation of vectors as little more than a by-product in a system which – he believed – held the key to all mathematics, all science, and perhaps all philosophy.

After fifteen years of intense thought, he published his first ideas on quaternions in 1843. He had still more than twenty years to live, and all these were overcast by his resentment of the world's comparative neglect of his great vision. He simply worked on – eating little and drinking deep, despairing of men but still trusting in God – until he died of gout at the age of sixty.

Yet the world was, for once, right; for quaternions never quite showed the way to that universal mathematics which their inventor had promised for them. And in the field of vector analysis itself – of the mathematics, that is, which applies to the combination and interaction of various types of force – humbler men produced easy-to-handle methods which were to prove adequate until all such approaches became embodied in the 'tensor calculus' of today – a technique, conceived by Riemann and using the ideas of matrix algebra, which itself, ironically, owes much to the work of the unhappy Irishman.

The mathematics of vector analysis and even of quaternions (though not of matrices and tensors) can be approached without more than an elementary background knowledge. Here, though, we can single out only two points for special note. One is that a system of vector-multiplication was evolved which satisfied the mathematician by being self-consistent and the physicist by explaining the experimental behaviour of heat-flows and wave-actions, and that it turned out that in such a system the product of a force working up or down this page and one working across would be one working into or out of it. The result will perhaps not be too surprising to anyone who remembers either the significance of '*i*' or the 'three finger' rules of elementary electro-magnetism. What *was* surprising – and so much so that it took Hamilton years to appreciate the fact – was that in vector multiplication A times B did not necessarily equal B times A. In fact, in the particular problem on which the Irishman was concentrating, $AB = -BA$: he realised this during a walk beside the Royal Canal in Dublin, and celebrated by immediately scratching the formula on a handy stone bridge.

Insult to common sense as this kind of behaviour may seem, it is no more remarkable in a world of rotations than the useful appearance of the 'imaginary' $\sqrt{-1}$. Supposing we take a cut-out triangle, for instance, and define A as the *operation* of turning it 90° clockwise, B as that of 'flipping' it over on whatever may be its base at the time, and $\times$ as defining the order. The diagram opposite shows a pair of sequences in which A $\times$ B is certainly not the same as B $\times$ A.

Although Hamilton was looking for something quite different, we have here another link with the new algebraic ideas which were developed towards the middle of the last century. It will be better, though, to group these together in the next chapter, and for the moment to look briefly at one or two other themes which engaged algebra just after Gauss's time.

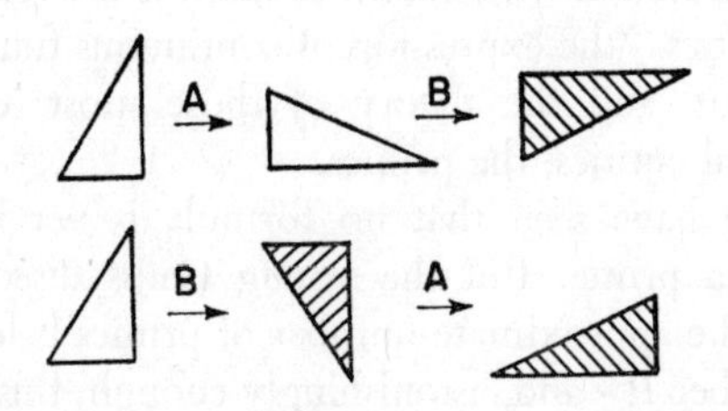

The long-established quest for precise solutions to equations continued, for instance, even though it largely lost touch with the practical needs of the scientists and engineers who had their own trial-and-error (or 'relaxation') methods for solving – say – equations of the sixth order. Another traditional investigation was that which explored the nature of numbers such as *e*; and here it was eventually proved, after some very tough thinking, that such numbers could never be extracted from finite equations and were at least as different from the $\sqrt{2}$ type of irrationals as these were from ordinary fractions. Following a suggestion which went back to Newton, *e* and its fellows were now christened 'transcendental' numbers.

It had, of course, long been suspected that π was of this type, but the search for a strict proof of the fact did not end until 1882. When it came it also ended the centuries of endeavour which had been spent on searching for a means of 'squaring the circle' by Euclidean methods. For geometrical constructions carried out with a ruler and compasses could never represent anything more complicated than the solving of certain quadratic equations.

In one sense, then, 'analysis' moved further away from traditional algebra. But such distinctions are largely man-made, and at the same time some remarkable links were formed between two studies which seem to stand at the very extremes of mathematics – the expression of continuous functions such as logarithms, and the theory of those most 'discrete' of mathematical entities, the primes.

Thus, we have seen that no formula is yet known for identifying a prime. But the young Gauss discovered one expressing the approximate number of primes below a given (large) number P – and, astonishingly enough, this was given by $P/\log_e P$. It has recently been shown that other types of number obey the same distribution, so that some mathematicians now suspect that the mystery of primes is less profound than has been thought for 2,500 years. But whatever the outcome, number theory cannot today be pursued without using the tools of 'analysis' as well as of pure algebra.

Finally, we must mention a few of the great names in nineteenth-century algebra. First in time – and second to none in importance – of the generation which followed Gauss was Cauchy, a mathematical descendant of Laplace and Lagrange. Alongside his work as a military engineer (for Napoleon had put him in charge of the planning of an invasion of England), the young Cauchy continued that revolution of 'rigour' in mathematics which we shall consider in the next chapter; and when he became free to devote all his time to creative work he threw out ideas so fast that he swamped the learned publications and had to set up his own journal.

Much of this work, too, was of great importance. Thus, in every division of their subject, mathematicians like to regard one theorem – however artificially – as 'fundamental': the fundamental theorem of arithmetic, for instance, states that a number can be factorised in only one way. Cauchy proved what is often called the fundamental theorem of algebra; and it is unfortunate for his reputation that his greatness of intellect was not matched by an equal largeness of heart, for his passionate piety and loyalty were offset by a certain meanness and carelessness in more human affairs.

A generation later French algebra was represented by Hermite, who proved the transcendence of *e*, lived into the present century, and had as a pupil Henri Poincaré. One of the last men to make important advances in every division of mathematics, Poincaré also carried on the work of Laplace in 'celestial mechanics' and became a talented populariser of science and mathematics. His studies, based on questionnaires sent to his colleagues, as to how a creative worker forms his ideas are still very much worth reading in a field where too little work has been done. And yet if one looks up 'Poincaré' in an encyclopaedia one is more likely to find Henri's brother, a rather routine politician.

We shall meet with one more important French mathematician later: others again, like the Simeon Poisson who discovered his great talent for mathematical physics only through becoming intrigued by a puzzle, we must pass by. It is not surprising, though, that under the influence of Gauss Germany awoke to the possibilities of the new algebra too. Names which are still important in the mathematics of today, for instance, include those of Bessel, Dirichlet, Jacobi (a particularly inventive manipulator of symbols) and Karl Weierstrass. If we give more space to the last of these than to the others, it is partly because two aspects of his work can be mentioned in non-technical form and partly through the interest of his life and personality.

Weierstrass's attention was largely devoted to a subject of

great practical importance and one which, perhaps, we have here under-stressed – the solution of those systems of simultaneous equations involving calculus functions which had, since Newton's time, cropped up in the study of numerous physical problems. (Thus, the 'three-body' problem in gravitation is essentially one of solving simultaneous differential equations.) But it is also of interest that he showed how a good deal of 'analysis' depended on the properties of the integers, and so provided a complement to Gauss's introduction of the mathematics of the continuous into pure arithmetic.

As a man, Weierstrass was one of the most lovable characters in the history of his subject, a big bear of a Rhinelander fond of his food and his friends. (So too was his contemporary Kummer, an arithmetician who carried Gauss's work into strange new realms.) But under this genial surface Weierstrass had at times found life hard. His father had dominated his early years and insisted on his studying law, for instance, and the son disliked this subject so much that he spent his student days in following the traditional German college sports of fencing and beer-drinking and so failed his degree. Fortunately, though, he was given a second chance, and then studied successfully under an eccentric mathematician whose course was far too advanced for his ordinary students. After the first lecture the only member of the audience was Weierstrass, and the course continued in a situation providing a perfect example of what is mathematically known as a 'one-to-one reciprocal correspondence.'

Soon, however, Weierstrass had to take a job as a master in an ordinary secondary school. He was a brilliant teacher who loved – and was loved by – his pupils at every level: later, for instance, he was to move heaven and earth to find a college place for a highly-talented Russian student whose only disqualification – in the middle of the last century – was that she was not only a girl but an outstandingly beautiful one. But these were years of 'infinite emptiness and boredom' when Weierstrass could relieve the tedium of his days only by long nights of creative work. One morning when he did not appear at school, for example, his headmaster went round to his house and found Weierstrass working behind drawn curtains, quite unaware that more than twelve hours had passed and daylight had come since he had begun his calculating.

Part of the trouble was that Weierstrass was so poor that he not only could not marry but could not afford the postage on the letters necessary to keep up a mathematical correspondence. At last though, when he was forty, he had an article published. And it is pleasant to record that thereafter he became steadily more widely-known, more prosperous – and more popular – until his death more than another forty years on.

Since we are now grouping mathematicians by their nationalities, it is a good moment to mention that this period also saw the re-entry of Britain as a great mathematical nation after the long post-Newtonian sleep. One major development was, in fact, to become largely a British speciality. But we should not overlook the more general work in algebra and 'analysis' of two men who were born in London – though one, Cayley, was part-Russian and the other, Sylvester *né* Joseph, was wholly Jewish.

By a remarkable coincidence, both these had graduated from Cambridge as the *second*-best mathematicians of their years, just as Kelvin and Clark Maxwell were to do a little later. In every case the candidates who passed first are now forgotten; but the fact that the roll of 'second wranglers' between 1837 and 1852 includes four men of genius is worth remembering in days when all examination results are regarded as suspect.

With six years between them, Sylvester and Cayley did not meet until both were practising – successfully but rather unhappily – as lawyers. (Earlier Sylvester had taught mathematics privately, one of his pupils being a young lady called Florence Nightingale.) But then the two immediately formed a friendship which was based partly on their common field of interest – which, in essence, was the 'invariance' of mathematical relationships in the sense suggested earlier in this book – and partly on the fact that the rather stolid Arthur Cayley formed a perfect balance to J. J. Sylvester – a much more mercurial character, a part-time poet and musician whose mathematical memory was so bad that he sometimes refused to believe what he had proved the day before. This friendship was uninterrupted even when, later in his life, Sylvester became the first professor at Johns Hopkins university in America – and hence, a century ago, one of the earlier Europeans to help found a new tradition in mathematics.

Finally, before closing this chapter, we must flash back in time to mention two algebraists whose lives were woven about a thread of tragedy. Of these, Niels Abel had begun his brief and poverty-stricken life in Norway at the very dawn of the century. As soon as his talents were clear his government raised a special fund to allow him to travel to Berlin and Paris; but he found the French mathematicians insular and inhospitable, and his only real supporter was a German publisher.

So Abel returned to Norway, to the support of his widowed mother and six young brothers and sisters, to the discovery that he was suffering from tuberculosis, and to his death at twenty-eight. His life-story is strangely similar to that of the poet Keats, even down to the one hopeless love affair. But a writer could reject as too obvious the crowning irony by which, two days after Abel died, the news came through that he had at last been promoted to a well-paid and long-hoped-for professorship.

Perhaps Abel's greatest single discovery was concerned with the solution of fifth-degree equations: this subject had challenged mathematicians for a century, but the young man's proof that a general solution was impossible settled the matter for ever. Otherwise, his work was largely in the highly-specialised field of exploring the multi-periodic functions referred to earlier. At that time these seemed to hold the unifying secret of all mathematics, but today Abel is revered more as another pioneer of the 'rigorous' approach. It is also a tribute to his lead that within a century Norway had produced other great algebraists – including Sophus Lie, the most distinguished of a talented family.

Nine years after Abel, a boy was born whose life was to be even shorter and more dramatic. For Evaniste Galois, the son of a rebellious French country mayor who eventually committed suicide, excelled his father in wild radicalism. If Abel was the Keats of mathematics, then Galois was a blend of Byron, Shelley and Chatterton – though he also foreshadows the young 'rebel without a cause' whose heyday came a century and a half later. Admittedly he lived in an age of revolutions and romanticism. But even his fellow-students could not follow him through the years when his passion for creative mathematics stampeded on beside a formless hatred for the 'malicious social organisation' which he saw as the enemy of all genius – and especially his own.

Galois certainly appears to have had more than his share of sheer bad luck. But he was also (in the language of today) emotionally accident-prone and possessed by a drive towards self-destruction. It is never a good idea to throw blackboard-dusters at one's examiners, and before Galois was nineteen it was clear that no university could hold him.

For a year or so, as the outside world saw it, Galois played at revolutionary politics; and eventually he achieved the martyrdom he secretly longed for in the shape of a light prison sentence. He was released from this a few months before his twenty-first birthday at the end of May 1832; and just what happened that day is one of the mysteries of mathematical history. Its events appear to have involved a prostitute with whom the rather priggish Galois had become compromised, but it also seems that there was a political angle to them. Whatever the excuse, Galois found himself challenged to a quite pointless duel.

It could well have been a bloodless one, too, like the majority of the then-fashionable student engagements: Weierestrass was to fight dozens such without receiving a scratch. But Galois was resolved on self-destruction. He had already drawn up his will, which like that of the poet Villon was a mixture of ordinary 'testamentary dispositions' and bequests of ideas. Now, in one feverish night, he scribbled down outlines of the mathematical thinking which had occupied him in those years between fifteen and twenty when he had seemed only to be playing at politics – thinking which included proofs of the impossibility of two other constructions which had troubled the Greeks (the doubling of a cube and the trisection of an angle) and the formation of ideas which, like those of Abel, remain important today. He broke off only to write notes to his few friends, and ended every scrawl 'There is no time . . . No time.'

At dawn Galois went out to die, and succeeded (which is rather hard in duelling) in catching a bullet in his guts at twenty-five yards. Even this might not have been serious were it not for the most inexplicable and Galois-like feature of the whole tragic farce, which was that everybody else promptly quitted the 'field of honour'. The boy bled and rotted internally for a few hours until a passer-by dragged him to hospital; but by then it was too late. They threw his body into a pauper's ditch.

As counter-balance to a story without parallel in the biographies of the world's men of genius, we should end with a word on the lives of the nineteenth-century mathematicians in general. They had, of course, such personal characteristics as Hermite's lameness and inability to pass examinations, or Poincaré's clumsiness, acute visual memory and mild kleptomania. But it is curiously easy to draw a kind of composite, 'Identikit' portrait of the *typical* mathematician of the age.

He may have come, for instance, from the huge (and undistinguished) family of a Protestant parson or relative of a Roman Catholic priest: he himself probably lived a life of traditional piety, and from a contented marriage produced far more (and undistinguished) children than he could support. He was gentle, peaceful and dignified even in his professional dealings, rarely engaging in the public controversies of – say – the chemists of the time. He lived into his eighties and (contrary to the legend that the mathematician's creative life is short) was mentally acute until the end. In France, Germany or Italy he saw revolutions and counter-revolutions rumbling round him and had a critical sympathy with the new ideal of equality; but he was never misled into confusing mob violence with democratic will. Humble as his own social background may have been, his intellect made him a believer in selectivity, human quality and an aristocracy of the mind.

He was, in fact, a natural conservative. Yet in his work he might well be laying a hatchet at the roots of a tree of thought which had grown unpruned for over 2,000 years.

7
Yesterday and Today

Throughout the history of mathematics two themes appear in tension if not actually at war. There is the desire for *precision*, for giving more accurate meanings than do our intuitions to such terms as 'point' or 'number'; and some philosophers of the subject regard this movement as typical of the mathematical approach. But equally important has been the tendency for mathematicians to 'bash on regardless' – to take those short cuts of thought which have sometimes proved very fertile, have sometimes (as Poincaré put it) led to mathematics becoming 'the art of giving the same names to different things' – and have sometimes concealed slips in basic reasoning.

For the early Greeks, logic had come foremost: if they could not precisely define an entity, then it had no place in their thinking. But this approach soon exhausted itself, and from Archimedes on there was an increasing trend for mathematics to forget even its great rule of self-consistency in the race for new ideas. We have seen, for instance, how Leibniz' infinitesimals were 'there but not there', constant yet variable; and d'Alembert had even suggested that a mathematician, like a religious convert, should close his eyes to inconsistencies in the hope that illumination would come. For what seemed to matter most in the eighteenth century was that mathematics such as was embodied in the calculus produced magnificent results.

But we have also seen how men like the Bernoullis and Euler began a movement back towards Greek standards of clarity, and how this was consolidated by the three great pioneers Gauss, Cauchy and Abel. This step was, indeed, enforced by the working demands of algebra: for instance, it had become clear that the ideas of 'convergence', 'limits' and so on used in the handling of series were leading to contradictions and in need of much closer inspection. At the same time, too, a dream was taking shape which had a parallel in the eighteenth-century belief that it should be possible to invent a universal language of ideas. So far as words went the dream faded; but the story of the quest for a more generalised and consistent symbolism forms the most important theme in the history of pure mathematics for the century after (say) 1820.

Before we start to trace this story, we should define what was at stake. Though the goal was not clearly seen at first, it was no less than that of rebuilding the entire structure of mathematics (elementary logistics, the arithmetic of integers, algebra and 'analysis', Cartesian geometry, the calculus) on a basis as firm as Euclid had taken for his limited geometry – on a *firmer* basis, indeed, for this time there must be no hidden assumptions. Everything must be developed logically from the smallest possible collection of data – ideally, perhaps, from just 'unity' and a handful of rules of combination – before men had a right to make even so simple a statement as that $\frac{1}{2} \times \frac{1}{2} = \frac{1}{4}$.

It should now be clear why in the last chapter we hinted at a coming revolution in mathematics; and when this movement gathered strength it became a suitable partner to the other great upheavals which the later nineteenth century witnessed in the sciences, the arts and thinking generally. But it all began quietly enough, and largely as a result of the work of three Englishmen who took advantage of their country's comparative isolation to develop a new approach instead of rejoining the continental main-stream as had their contemporaries Sylvester and Cayley. It is hence ironic that one of the predecessors of this movement towards a logical mathematics had been the man partly responsible for the great divorce between Britain and the continent, Leibniz.

The first hint of this approach came from another pair of Cambridge 'twins' – Charles Babbage and Augustus De Morgan. De Morgan, whose sons became distinguished in the world of Victorian art, had his own eccentricities; but Babbage, who followed him some fifteen years later, was one of the oddest figures ever to adorn the sciences. Whether attempting to skate on water, to survive being roasted in an oven or to raise the devil, he was always experimenting to his personal danger; and perhaps the most remarkable fact of all about him is that Babbage eventually died of natural causes at the ripe age of seventy-nine.

Because we shall be considering connections between mathematics and logic, Babbage's most famous work is not irrelevant here. Ever since Raymond Lully had foreseen the possibility back in the thirteenth century, men had attempted to construct simple machines which should illustrate the premisses and syllogisms of the traditional verbal logic of Aristotle. These devices were rarely worth the labour put into them; but the fact that there was nothing absurd in the idea is shown by the geometrical illustrations of such logical ideas as the relationships between classes which were invented by another member of this school, Venn, and which are still found useful in teaching.

The figure opposite, for instance, illustrates the statement that 'some cats are black'. And since pure logic, like pure mathematics, is not concerned with mere *facts*, similar diagrams could be drawn for 'no cats are black', or 'all cats are black', or 'all black things are cats'.

Although Babbage himself played with such ideas, his famous 'folly' was not a simple logical model: it was intended to do useful work, the calculation (and direct printing) of astronomical tables. But it was not a mere calculator of the Pascal–Leibniz type either. It was a true *computer* in the modern sense, a machine which could 'think' to the extent of using repeated operations to solve equations – for example, by

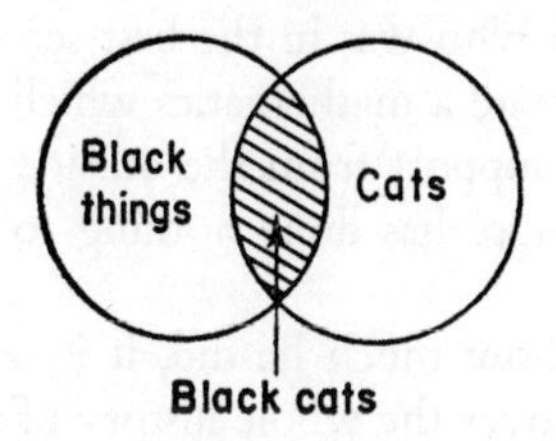

taking advantage of the calculus of finite differences. An engine 'as big as a barrel' and designed to be driven by steam, it never quite *worked;* but its failings were purely mechanical ones, and in the mathematical and logical ideas which he incorporated Babbage was a century ahead of his time.

Babbage and De Morgan were men of comparative leisure and wealth, but neither made so great an advance as did George Boole. Born like Weierstrass in the historic year of 1815, Boole was the son of a provincial shop-keeper, was largely self-educated, and did not discover his full talents until as a junior school-teacher he found himself struggling with the very poor textbooks of the period. He then suffered fifteen years of grind, struggling to support his ageing parents; but at the same time he formed ideas on the teaching of elementary mathematics which were not to be incorporated into school courses until over a century had passed, and also slowly built up a reputation by his creative work. At last, like Weierstrass again, he received his reward; and the man who had never taken a degree was asked to step straight into a newly-established professorship in Ireland.

Boole was then thirty-five and had only fifteen years left to live. But (thanks partly to a happy marriage) these were rich years. And though the importance of his work on 'the laws of thought' was not at first appreciated, it was being said fifty years later that before Boole there was no truly 'pure' algebra and that this man who was in the best sense 'self-made' was also the first to create a mathematics which stood on its own feet and without support from the outside world. A further half-century or more has done nothing to diminish Boole's reputation.

To appreciate how much he did, it is necessary to take a backward glance over the whole history of algebra. We have seen how, ever since Babylonian days, the same methods of manipulating symbols had been applied to two different ends – the determination of unknown quantities through the solving of equations, and the establishing of the generalisations which enable us to compress an *infinite* range of data (for instance, $1+1 = 2, 2+2 = 4, 3+3 = 6 \ldots$) into a single formula ($\mathrm{A}+\mathrm{A} = 2\mathrm{A}$). Equation-solving had for a thousand years seemed by far the more useful and glamorous field. But it was now becoming suspected that the basic algebraic identities deserved more attention than they had yet received.

Until almost 1800 it had been assumed that the only reasonable ways for any mathematical object to behave – say, in addition or multiplication – were those which ordinary numbers seemed to obey in the 'real' world. Typical of these rules, which as usually taught seem quite arbitrary, is that $A+(B+C)$ always equals $(A+B)+C$, but that $A\times(B+C)$ usually does *not* equal $(A\times B)+C$. But even for ordinary-looking numbers in the modular arithmetics of Gauss the 'commonsense' rules needed some modification; and it was becoming clear that anybody needing a reasonable mathematics of vectors, of the ordered groups of numbers called determinants and matrices which were now finding a place in co-ordinate geometry, or even of ordinary rotations, would have to swallow the principle that for some operations AB regularly equalled $-BA$.

Boole himself began by adding another non-numerical type of entity to these. Suppose we call the set of cats C and the set of black things B. The set of black cats will obviously be reached by choosing the elements common to both these, and is often called the *intersection* of B with C and written $B \cap C$. But we can also combine the two sets by simply forming a *union* of B and C, written $B \cup C$ and interpreted here as 'the set of all objects which are *either* cats *or* black' (including, of course, those which are both).

This second operation looks rather like ordinary addition. Does intersection, then, in any way resemble multiplication? The set of 'black cats' is certainly smaller than those of either 'cats' or 'black things' – but this is just what one would expect when multiplying *fractions*, and both are small when compared to the 'universal set' of all possible objects. And in fact Boole was able to build up an entire algebra from the logic of such sets, an algebra in which – for example – the fact that $(-1) \times (-1) = +1$ appeared as a parallel to the 'two negatives make a positive' rule of ordinary grammar.

Furthermore, another and even more Aristotelian aspect of logic fell to the same treatment, the 'calculus' of propositions and implications. A sentence such as; 'I watch TV on Saturdays if, and only if, it rains: I watched last Saturday: what was the weather like then?' illustrates this type of analysis, which seems trivial until extended into the long chains of reasoning which were the delight of another of those eccentric Victorians (though this time an Oxford man) who brought a new vitality to a subject almost dead since the Middle Ages. This was the Rev. C. L. Dodgson; and if the reader is by now feeling that he has slipped out of the world of mathematics into that of *Alice in Wonderland* he is quite right. For Dodgson is far better known as 'Lewis Carroll'.

For a hundred years or so Anglo-Saxon mathematicians, logicians, physicists and philosophers have treasured Carroll's children's books as analogies of their thinking: *i*, for instance, was historically a very slithy tove, an infinitesimal is like the grin of a Cheshire cat, and the mathematician must still make sure that when he is hunting a convergent snark called $\Sigma f(x)$ he does not catch a divergent boojum instead. But what is clear is that, in the middle of the last century, there was an increasing number of algebras to choose from, depending on whether one was dealing with quaternions, black cats or ordinary numbers. Each used its own symbols, with $\wedge$ and $\otimes$ as well as $\cap$, for instance, indicating a multiplying analogue. Yet in each there were self-consistent rules for combining the various entities, and – more surprisingly – there was a strong family resemblance between these various algebras.

So it was that mathematicians came to realise that the way to set their world straight was to stand it on its head. Instead of regarding as an irritating minor anomaly the fact that turning-and-twisting a triangle was not the same as twisting-and-turning it, they decided to give this inequality just as much weight as the fact that 3×2 *did* equal 2×3. Both were possible patterns of behaviour for mathematical objects.

Before going on, we should stress that this was no sudden revelation. Since the start of the nineteenth century, at least, the 'rigourists' had been not merely questioning the foundations of traditional algebra but doing what political revolutionaries usually fail to do, which is to suggest a workable alternative. We have seen how the quest for a new discipline in geometry had led first to the destruction of one of Euclid's axioms but afterwards to a greater freedom for mathematicians to construct 'possible worlds' which might or might not mirror real ones: now a similar freedom was to be brought to algebra as the examination of the mathematics of logic modulated into that of the logic of mathematics. But, whatever these new mathematics might look like to the outsider, there was nothing anarchic about them. They had to be *more* rigorous, *more* tightly-argued, *more* free from vague intuitions and concealed self-contradictions, than the old ones.

A mass of such work, begun by Boole, was summed up towards the end of the century by the Italian Guiseppe Peano who took over an idea of Sylvester's and postulated a set of entities characterised by their behaviour under certain operations. (The ugly wording here is inevitable, for we are using words to represent very generalised ideas: thus, we have no right as yet to call these operations 'addition' and 'multiplication'.) Peano found that this behaviour was particularly clearly shown up by the answers to three questions:

(1) Was the result of any such operation in the original set? For instance, the sum or product of any two positive integers is itself a positive integer; but subtraction can lead to a negative integer, and division to a number which is not an integer at all.

(2) In what ways were the operations reversible, or 'spreadable' over groups of entities?

(3) Was there an element in the set whose use left every other element unchanged, as the addition of 0 or multiplying by 1 does with numbers? A similar question could be asked about the presence of 'inverses' like negatives and reciprocals.

Some readers may wonder why so much space has been given to an apparently very technical matter; but in fact Peano's criteria (which could finally be reduced to a mere four in number) proved so fundamental that they are nowadays taught in some schools as an *introduction* to formal mathematics. Whether or not this is wise, the subject is certainly not intrinsically difficult: the difficulty is rather in appreciating just how deeply such rules probe. But anyone who has followed this book should be able to see that they present us with the possibility of a dozen different algebras, depending on whether we insist that the entities we are working with obey (say) law 2c but leave them free to disregard clause 7a(i), or *vice versa*. The ordinary numbers of arithmetic respect one collection of rules, and so ordinary algebra had grown up on the same model. But if Gaussian number systems, Boolean logics or Hamiltonian quaternions called for slightly different principles, there was no cause for alarm.

This is the meaning of such statements as that Boole discovered, and Peano codified, 'pure' or 'abstract' algebra: after their revolution, mathematicians could say they were working with 'an Abelian group', 'a Galois arithmetic' or 'a Lie algebra' much as it suited them. But before we pass on to the final phase of this re-thinking, there are several ideas which must delay us for a number of pages.

One of these is that the concept of sets, groups or classes, which we first encounted back in Babylonian times, was fundamental to this new mathematics. Several of the men mentioned in the last chapter (such as Hamilton, Cauchy and Cayley) had been responsible for building up an algebra of sets which produced some curious results – for instance, that the study of the possible positions taken up by a rotating icosahedron (a twenty-sided solid, the most complex of the regular bodies known to Pythagoras) led to a set similar in 'shape' to that of the solutions of the troublesome fifth-order equation. Out of the many implications of a theme which has for more than a century proved influential and productive we can here select two for special attention.

Firstly, then, most of the collections which we meet with have two things in common. Their members may be as small as electrons but are still discrete or naturally separate from each other, and as plentiful as all the electrons in the universe but are still finite in number. Mathematically, though, we can conceive of sets whose components are *continuous* (all the points on a line one inch long), *infinite* (all the integers), or both (all the points on an infinitely-long line). Can we say anything meaningful about such sets?

Ever since Zeno had baffled the Greeks with his paradoxes, mathematicians had kept clear of head-on collisions with infinity: indeed, we have seen them getting into trouble in their enforced dealings with the infinitely small. But the subject could not be avoided for ever. First an Austrian priest intrigued Weierstrass with some new ideas on it. Then it was the turn of Dedekind – a mathematician remarkable for more than his life-span (he was born when Legendre was still alive and he died, the last pupil of Gauss, in 1916), since he showed how certain algebraic problems could be brought into line with arithmetic if an entirely new type of number was used. And finally, around 1875, a man of very unusual intellect who had earlier done much to clarify the idea of sets in general devoted himself to seeing if any consistent ideas could be formed by experimenting with infinity.

This man was Georg Cantor. Like many of the great mathematicians who worked between about 1830 and 1930, Cantor could be described as a German Jew; but in fact both his national background and the religious leanings of his family were very mixed, and it may be relevant to his choice of subject that he had steeped himself in the thought of those medieval Christian philosophers who had asked how many angels could dance on a needle's point. Cantor's line of thought, remote though it is from all practical mathematics, is particularly worth following here for three reasons – because it is largely self-contained, because one or two of its proofs are as simple and beautiful as anything known to Euclid, and because it affords an unusual chance to appreciate that astonishing inventiveness – greater, perhaps, than that of any artist – which is the hall-mark of the creative mathematician.

Cantor's first principle was that two groups could be said to be numerically similar, *even when neither could be counted*, if every member of one could be paired-off with a member of the other. The term used for this pairing – a term which we light-heartedly introduced when mentioning the lecturer with only a single student – is 'forming a one-to-one correspondence'; but the idea was clear to Galileo and probably goes back beyond number itself to pre-Babylonian peoples checking-up on their sheep.

Now we can pair-off – for example – the members of the entire set of positive integers with their 'doubles' ((1,2), (2,4), (3,6)...), with their squares ((1,1), (2,4), (3,9)...), or with any other ordered sub-set of integers we choose. It is not even necessary for the members of the second set to be straightforwardly linked to the integer series: if we can show that they are being counted-off in some systematic way with no gaps left (as we can with the primes, for example), then we have established our correspondence.

So we can arrive at any number of double series like that shown opposite for the odd numbers: all will extend to infinity (as the dots indicate) with every number having its partner. And yet in this case the numbers in the lower line are 'obviously' only half as plentiful as those in the upper one, since we have left out the even integers. And this leads to the paradox that in the arithmetic of the infinite a whole may be no larger than a part of it.

$$\begin{array}{cccccc} 1 & 2 & 3 & 4 & 5 & \dots \\ \updownarrow & \updownarrow & \updownarrow & \updownarrow & \updownarrow & \\ 1 & 3 & 5 & 7 & 9 & \dots \end{array}$$

Faced with such an apparent contradiction of the common-sense, Aristotelian logic on which men had built traditional mathematics, Cantor did not give up. Instead, he boldly *defined* an infinite set as one in which a sub-set could be as great as the whole. And any ordered or *denumerable* collection of entities, he said, was just as 'infinite' as any other.

Can we find other collections of this sort? The criterion is that we must be able to write down its members in such a way that we are sure we cannot slip any more into the order; and since we shall be using the methods of mathematics, our proofs will probably be restricted to mathematical objects. First, Cantor showed that the set of all rational fractions made up such a denumerable infinity (i.e. could be paired-off, one-by-one, with the integers 1, 2, 3, etc.): the proof is so simple and elegant that it is found in almost every popular book on mathematics. Then he did the same for the set of *algebraic* numbers or possible solutions to equations – and such is the nature of the subject that here the proof calls for much more advanced treatment.

Does this mean, then, that we can now count off any imaginable group of numbers and label its members, 1, 2, 3, 4...? Not quite – for we have not yet considered the 'transcendental' irrationals such as e and π, those which can only be expressed as series. Again, a proof applying to the transcendentals *alone* belongs in the loftiest realms of mathematics. But we can approach the matter another way and try to discover whether the entire 'number line' consists of denumerable points.

It is, incidentally, interesting here to see how the idea of a number-line (which we casually introduced on p. 67, and which was itself developed by Dedekind) sums up much of the history of mathematics; for beginning with positive integers we have followed history in adding negatives, rational fractions, algebraic numbers and then transcendentals before using *i* to take us into a number *plane*. But at the moment we are concerned with one of those proofs which – like that of the infinity of primes – appears impossible yet demands only the simplest of reasoning. It is a proof so Greek in form as well as spirit, so classic an example of the continuity of mathematics, that – familiar though it is in popular books – we must include here an abridgement of Cantor's demonstration that the totality of all numbers *cannot* be denumerated.

In fact, Cantor begins by narrowing the field to numbers between 0 and 1. He then asks anyone who thinks that these

	• • • • • • • • •
π/10	0·3 1 4 1 6 • • •
8/25	0·3 2 0 0 0 • • •
1/3	0·3 3 3 3 3 • • •
$\sqrt{2}/4$	0·3 5 4 5 3 • • •
	• • • • • • • • •

can be completely 'catalogued' to bring him, not the catalogue itself (which would be infinitely long), but some rule by which this can be produced: note how the mathematician, like a Judoka, prepares for victory by bowing submissively to his opponent. A sample from a list prepared according to such a system, for instance, might be like the one shown here in decimal form. The left-hand column demonstrates that our sample includes every kind of 'real' (or non-*i*) number we know of: the dots mean that every individual number is to be continued 'to infinity' (with zeros if it terminates) and that the list itself, though supposedly denumerable, will be infinitely long.

Now, Cantor says, consider a number like that enclosed in the diagonal box, 0·3235..., which has been made up by systematically selecting one digit from each of the other numbers. We can as systematically 'mutilate' this by adding 1 to every digit in it (putting $9+1 = 0$) so as to produce yet another number, in this case 0·4346... This must differ from the first number on the original list in (at least) its first decimal place, from the second one in the second place – and from the 'infinitieth' one in the 'infinitieth' place. It is a new number which could not have appeared on the list we were promised, and so the cataloguing rule was inadequate.

Since we can defeat any such rule by the same graceful *reductio ad absurdum*, the infinity of all numbers must (thanks to the intervention of the transcendentals) be of a different 'power' from the mere infinity of integers. Cantor indeed proved that this new, *continuous* infinity was itself matchable with that of the points on a line. It is easy to show that there are the same number of points in any line, whatever its length, harder to show that just the same number of points would fill all space, but *possible* to show that even higher orders of infinity can be constructed.

This subject of 'transfinite' numbers is a fascinating one. But we should not leave it without mentioning that there have been mathematicians even more rigorous than Cantor himself who have denied that anything meaningful can be built up on the idea of infinity. For instance, Cantor's near-contemporary Kronecker – another Jew with a keen interest in theology, but one who also had a passion for the economy of ideas – felt so strongly on the subject that he began the most bitter personal battle in the history of a subject generally free of such blemishes. Earlier, Kronecker had baited Weierstrass over a similar controversy (and 'baited' is an appropriate word, for whilst we have compared Weierstrass to a shambling great bear, Kronecker was a dapper little terrier only five feet high); but those differences had been put aside with the good humour which Socrates seems to have met with in *his* hair-splittings. This later attack, though, had the effect of driving the sensitive Cantor into a mental hospital.

Few readers will want to risk following him by attempting to think too hard about the apparently unthinkable. But creatures as strange as the transfinite numbers will be found prowling in another field where the concept of sets has proved very influential. When we last looked at 'analysis situs' it was largely definable as a geometry whose figures were drawn on stretchable rubber sheets, and this still forms one aspect of this branch of mathematics. It is only a few years, for instance, since the problem of 'everting' a hollow sphere was solved.

To evert a body means to turn it inside out like a glove. Obviously we cannot do this to a real-life sphere like a tennis ball; but mathematical spheres are not made of rubber but rather of collections of points which can pass through each other without trouble. And on those terms there at first seems no difficulty about pushing one side of a sphere through the other.

There *is* a difficulty, though – the fact that we shall at the last stage be left with a looped ridge. Under the rules we cannot get rid of this; and hence, if a sphere *can* be everted, the method must be a less direct one. The approved procedure in fact involves torturing the sphere into a dozen stages of weird, saddle-like shapes whose sides pass through each other:

once done, the process can *just* be envisaged, but it could never have been invented by a man thinking in pictures rather than symbols. And in fact the French mathematician who worked out the method of eversion was blind . . .

'Analysis situs' today, however, is very little concerned with spheres or with anything else which can be visualised. It *is* concerned, as we hinted above, with groups of points in space; and because these sets can themselves represent algebraic functions of many kinds the subject has taken over a good deal of 'analysis', blended this with ideas from all the older geometries and (even more) from the theories developed by Boole and his successors, and finally emerged with the smart new name of 'topology'.

The word 'topology' means much the same as 'analysis situs', the study of places; but mathematicians are notoriously bad at their 'naming of parts', and we have just seen that topology is now much more concerned with abstractions than with pictures. It has become an increasingly popular branch of mathematics in the last thirty years, and today there are many universities with departments of topology where mathematicians hold perpetual debate or sit pondering like monks in cells built round concrete cloisters. But one will not often find them attempting to turn inner tubes inside-out or to tie 3-dimensional knots in 4-dimensional spaces.

Topology and set theory are in fact now very fashionable subjects, promising the kind of union of all mathematics which the study of bi-periodic functions did 200 years ago. Whether they will live up to the expectations which some have for them is another matter, though. For mathematics can have its gallant failures, as one is reminded when one returns to the approach which seemed to promise so great a unification towards the end of the last century.

We left mathematical logic at a stage where the decks had

been cleared for a major reconstruction. The tools for the job were to hand in the form of Boole's algebra of groups and propositions; now, around 1900, the mind to apply these tools emerged. This outstanding intellect was that of Bertrand Russell, who was mathematician, logician and philosopher in a Greek or renaissance fashion but who in his long decline unfortunately aimed at being a moralist and politician too. Inspired by a meeting with Peano which opened his eyes to ideas then new to him, Russell read all that was available on the very tough subject of mathematical logic in a single month. And in four more he produced, as the result of what he termed an 'intellectual honeymoon', the first draft of one of the great works of the human mind.

Called *Principia Mathematica*, this took as a starting-point the assumption that mathematics and logic are almost synonymous and that only through them can *any* certain knowledge be won: indeed, many of the mathematical difficulties which Russell and his collaborator A. N. Whitehead had to overcome can be presented as verbal paradoxes. One of them, for instance, concerns what we are to make of a slip of paper with the message 'The statement on the other side of this paper is false' written on one side and 'The statement on the other side of this paper is true' on the other – and it is not solved by twisting the paper into a one-sided Mobius band! Another such paradox deals with the dilemma of a bookseller who is asked for the complete novels of Jane Austen and does not know he should include an 'omnibus' volume, *The Complete Novels of Jane Austen,* as well as all the individual books.

These may sound like party conundrums; but the inability to resolve such inconsistencies can prove the link at which a whole chain of argument snaps. Thus, one such teaser was propounded by Russell to his opposite number on the continent, Gottlieb Frege, just as he was putting the finishing touches to a huge book of his own on the logic of classes – and forced Frege to the conclusion that there was a tiny inconsistency in his life's work which invalidated the whole scheme.

But there were also more purely mathematical difficulties in the attempt to justify, by strict reasoning from a handful of principles, that great structure of mathematics which we have watched growing up through the ages; and most of these clustered about a bridge which had seemed too weak for its traffic since the time of Pythagoras and Zeno, the bridge between the discrete and the continuous. There were, indeed, neo-Pythagoreans or pre-Babylonians like Kronecker who held that the only true 'numbers' were the positive integers useful for counting sheep, and that all the rest of the mathematical objects which we have met with in this book – negatives, fractions of all types, vectors and sets and so on – were constructions of the human mind, useful but so artificial that it was absurd to suppose that they had a nature to *be* explored.

This 'formalist' view was taken to the extreme by David Hilbert – the last of the Göttingen giants, and perhaps the last man able to be great in every aspect of our subject including its philosophy – who claimed that mathematics was no more than a game played by making marks on paper – marks which might accidentally have *applications* in real life but which were themselves meaningless. It also received an unexpected endorsement in the 1930s. Several times in this book we have stressed that the mathematician's first demand of his methods is that they should be *self-consistent,* and nowhere is this more important than when we are examining the logic of the subject. It was hence a major aim of Russell and his school to prove that the structure which they were shaping *was* consistent – and a considerable shock when an Austrian logician showed, by an extremely complex but unchallengeable chain of symbolic reasoning, that any such proof was itself impossible.

But long before this Russell himself had become disillusioned with his attempt to rebuild mathematics, and had turned to an even less hopeful attempt to rebuild human nature: this quest occupied more than half of a life which itself extended for nearly a century. We have met many examples of wasted talents in this book, ranging from the preoccupation of the seventeenth-century Protestants with fine points of theology to the self-destruction of Galois. But it may be that the loss from mathematics of Bertrand Russell is the most tragic such case of all.

Russell's great examination itself, however, was not wasted: very little in mathematics is. If it did nothing more (and in fact it did a great deal more), it gave new urgency to a question which had barely been asked since the days of Aristotle – the question 'What *is* mathematics?' It is clear that we cannot answer this here; for it has divided experts for at least the whole of the present century with no more progress being made than ever was with the fundamental dilemmas of philosophy and theology, and today seems, like them, to be shelved rather than resolved. But it should also be clear that the question is far thornier than any concerning the fourth dimension or the square root of minus one. And that is one reason why we have given so much space here to the logic of mathematics.

It would, though, be false to give the impression that pure mathematics in this century has been entirely concerned with self-examination. We have already mentioned some of the other themes which have absorbed it, and in addition traditional algebra and 'analysis' are not yet quite exhausted. Even in this bird's-eye view of the twentieth century in which we can mention a mere handful of mathematicians as representatives of their special subjects, for instance, we cannot omit the name of Ramanujan, a self-taught and tragically short-lived Indian who was invited to Cambridge after posting to Trinity the note-books which nobody in his own country could understand – and who is now recognised as the greatest algorist or symbol-manipulator since Euler. Ramanujan was equally brilliant as an arithmetician, so that it was said that he knew the peculiarities of the integers as a man knows his friends and met a taxi licence-number like a welcome new acquaintance.

Throughout the long history of mathematics, however, the pure side had needed to be fed with challenges from the applied aspect just as the latter has looked to the former for its techniques: in age after age, indeed, the division of attention between the two has kept remarkably close to a 50/50 ratio. And so, before this book reaches its close, we must look at a few of the ways in which mathematics has served needs as typical of the twentieth century as navigation was of the fifteenth, ballistics of the sixteenth, horology of the seventeenth, optics of the eighteenth, electrical theory of the nineteenth, or surveying and astronomy of almost every period.

The growth of technology in the present century has been immense; and it has given rise to a new type of mathematician, the man who concentrates almost *entirely* on practical applications. But it has not produced a corresponding need for new mathematics, for with a few exceptions even devices and techniques as novel as those of electronics have been able to draw on a great reservoir of ideas which had accumulated over three hundred years. What is characteristic of the present age is rather the way in which the higher mathematics have been tailored to such activities as industry, commerce, government and logistics in the modern sense. One distinguished

British mathematician, Ronald Fisher, even made an international reputation by applying advanced mathematics to agriculture.

As was suggested in the last chapter, a key-word here is 'statistics'. Thus, even such comparatively ancient mathematical curiosities as the way in which π can be estimated by dropping a needle at random, or the problem of the 'random walk' which concerns the probable displacement of a wandering man who keeps changing his direction, have found practical applications in recent years; and when the history of the twentieth century comes to be assessed its most influential man may well emerge as neither a dictator nor a scientist but the economist Maynard Keynes.

Keynes was an Englishman responsible both for original work in mathematics and for the idea that nations could and should use taxation, not primarily to raise revenue or even produce social change, but to control their economies. Forty years ago this was a revolutionary doctrine: today, for better or worse, it goes unchallenged in countries of every political complexion. But Keynes' first claim to fame was as the author of a book on statistical methods.

A technique complementary to statistics but a century younger is known as the 'theory of games'. Despite its name, this has little to do with the problems of Pascal's gamblers and is rather concerned with situations in which two or more 'opponents' draw on a pool of information which may change because of their actions. A typical example of the type of question it deals with can be found in a hard-fought game of 'Happy Families' in its final stage, where every player holds one member of the last 'family' but none knows where any other member is held. Is it better in such a set-up to have the first challenge (when you know nothing but your own card but, if you are lucky, are in a position to sweep the board) or the last (when you will have had the benefit of others' failures if you ever get a chance to use your knowledge)? Once a situation of this type is analysed, the result may prove just as applicable to a 'take-over' fight between companies, a struggle to win a building contract, a tactical decision on a battlefield ('If A fires first he has the advantage of surprise – but B knows where he is'), or even the razor's-edge politics of the handling of weapons which could annihilate the world. The theory of games would indeed be better termed the theory of conflicts.

In the comparatively peaceful world of industrial production, similar types of mathematics have been combined into 'operational research' to help form the type of link between information, machines and men which is summarised in the phrase that a certain equation forms a 'model' of a factory situation. One whole new department of applied mathematics, for instance, is concerned with such problems as choosing the best mixture for the 'Doggo' which we concocted a few pages back, given certain minimum and maximum figures for the prices of its ingredients plus the properties demanded in the end product and the need to make the best use of costly plant. This field of inquiry has led to increased attention being paid to the behaviour of functions which are not continuous in the sense which we have assumed earlier, but whose graphs display such breaks and sharp angles as are found in the tariffs of a transport organisation.

Another group of modern applications is to be met with in the 'theory of queueing'. Again, the phrase suggests one of the more limited of the hundred-odd branches which make up today's mathematics, though anyone who has sat in a car waiting to filter past some road-works may have been struck by the thought that a single such 'bottle-neck' can lead to delays over a very wide area. But just the same type of situation arises when there is a congested link in a telephone network and the 'queue' is an invisible one.

The theory of queueing, in fact, is just one aspect of the (mathematical) theory of *communications,* which itself can take the same shape whether the communicating medium is a high-speed railway, a 'Telstar' satellite or an order shouted in a factory. Since the idea of communication is today the centre of so much interest, there is a temptation here to be led still further on towards a discussion of whether mathematics can contribute – in other than its data-handling role – towards helping men with those social and environmental problems which, three-quarters of the way through the twentieth century, are beginning to appear a greater threat to civilised living than was generated by the invention of nuclear explosives. But the little space which we have left must be given to less speculative matters.

The mere mention of atomic power is a reminder that what appears the purest form of scientific inquiry can develop within a few decades into a great new technology, and that no firm line can today be drawn between pure and applied science. But all mathematics developed in the service of science is applied *mathematics*, and we have not yet mentioned the part which our subject has played in forming that so-called 'new physics' which is itself now nearly a century old.

It must now be a very brief mention, partly because some of the ideas embraced belong to the physical sciences and partly because little of the mathematics involved was in itself revolutionary. We can begin, though, by pointing out that there are two main aspects to the revolution in physics which particularly characterised the earlier years of the present century. For on the one hand this cast new light on the problems of cosmology and the nature of the universe, whilst on the other it affected men's views of the structure of the atom and of the innermost nature of matter.

That these two revolutions took place side-by-side, though, is not a coincidence; for they had in common such features as the relationship between matter and energy discovered by Albert Einstein. Einstein was a German Jew who received only a middle-level education before he joined the Swiss patent office for the sake of a quiet job which would provide him with enough leisure to work out a new approach to certain physical problems. His work on this was substantially completed (and, for all its revolutionary nature, confirmed by astronomical observations) before 1920, and the second half of Einstein's life – as of Russell's – was largely an anti-climax.

Although Einstein's name is one of the very great ones, he was not a mathematical physicist in the sense that Newton or Gauss were (since he barely set foot in a laboratory or observatory) and was not a pure mathematician either. His work was devoted partly to a re-examination of the fundamental ideas of mechanics begun by Ernst Mach and partly to re-interpreting such experiments as those of J. J. Thompson on the electron, and to building their results into the synthesis of electromagnetic, gravitational and other phenomena which Faraday had dreamed of a lifetime earlier. And in taking this approach he was perhaps the first, as well as the greatest, of another new type of scientist-mathematician.

Some of the physical implications of Einstein's ideas are described in other books in this series. From our present point of view, perhaps the most remarkable feature of it is the way in which earlier mathematicians, often working without any thought of applications, had provided this great synthesiser with his intellectual tools. A key idea of relativity physics, for instance, is the identification of one of the building-bricks of the universe as 'action' – energy multiplied by time, which is not an obvious quantity in the sense that power, or energy *divided* by time, is: this concept dates back to Euler. Another characteristic tool is the use of the tensor calculus to describe stresses in space more complex than those which can be handled by Newton's mechanics: this calculus, as we have seen, was developed by Riemann from the 'determinants' first used in the analysis of equations by Chinese mathematicians. And yet a third such borrowing was Einstein's use of the 'Lorentz transformation'.

When we introduced Cartesian ideas several chapters ago, we pointed out that one of the simpler ways in which a graph can be transformed is by rotating its axes about their origin. This will alter some of the characteristics of the equation which describes the curve, but leave others unvaried.

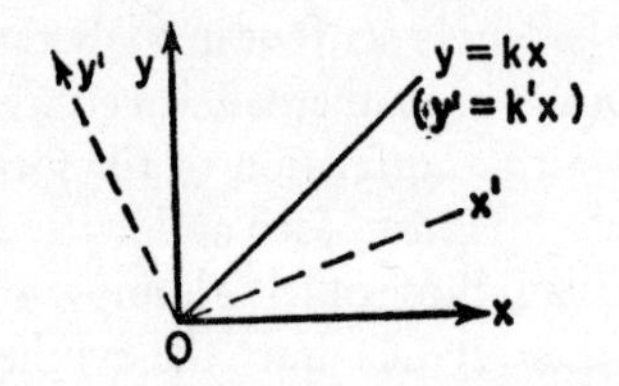

In the case shown, for instance, we would have to alter the 'k' in the equation of the straight line if referring it to the dotted axes rather than the original solid ones – but we would *not* have to introduce new constants or higher powers of x.

It will also be remembered that in many equations and graphs representing important physical situations – for instance, those typical of Newton's mechanics – x represents time, y distance and hence the slope of the curve a velocity. Now, what will a rotation of the axes of the graph imply in a situation of this type? About a hundred years ago, when scientists were becoming perplexed by certain problems of relative motion posed by their belief in a fixed 'aether' filling all space, it was appreciated that such a rotation could represent a movement by the observer studying the moving body.

The details of such transformations – such changes in 'frames of reference' composed of several dimensions – were worked out by a Dutchman, Hendrick Lorentz. His calculations can be found in a number of semi-popular books on relativity, and are well worth looking at by anyone prepared for some rather tough but not really 'advanced' mathematics – for they demand nothing beyond basic algebra. They show how an observer's own motion will affect his ideas about the movement, energy and hence mass of other bodies.

It can be seen from our velocity diagram, for instance, that a rotation of the axes will lead to a change in the ratio between the space and time components of the movement represented. Even in Lorentz' own equations time hence appeared as just another dimension of space, closely linked to (though not, of course, of the same nature as) the familiar three; and soon there emerged from this work of a century ago new relationships showing how the size, time-intervals, mass and similar characteristics of a moving body could only be standardised by reference to the 'absolute' speed of light. These equations themselves led to Einstein (and the Russian Hermann Minkovsky) constructing that unfamiliar universe of 'curved space-time' which almost every reputable scientist came to accept, nearly fifty years back, as presenting a truer picture of our actual world than that given by the more commonsensical ideas which had served for some 200 years after Newton.

We can give even less space here to the parallel development of the 'quantum mechanics' which showed how a number of problems concerning such matters as the emission of light from hot bodies could be resolved (as in Newton's physics they could *not*) by introducing two new ideas into calculations about events in the heart of the atom. One of these concepts was – as we have mentioned – that of the interconvertibility of matter and energy which is today involved in the generation of nuclear power. The other was that energy itself was 'discrete' or 'atomic' in its structure, so that radiations such as light should be regarded less as a steady flow of waves than as a series of packets of them, with each packet behaving in some ways like a solid particle.

Many outstanding men, drawn from several nations, contributed to that perfection of quantum theory which was the great achievement of physics in the 1920s and 1930s: important names here are those of Bohr, Bondi, de Broglie, Fermi, Heisenberg, Pauli, Planck and Schroedinger. All these were highly competent mathematicians, as any leading physicist today must be: *as* a mathematician, perhaps the greatest innovator was the Cambridge scientist Paul Dirac who first combined the ideas of relativity and of quantum mechanics. But – once again – the mathematics needed were rarely new in essence. The main problem was to bring to the heart of the atom the type of Gaussian probability-statistics approach which Clerk Maxwell and Kelvin had applied to atoms themselves in the last age of 'classical' physics, though an interesting sidelight was the way in which Hamilton's 'abstract' algebras proved valid for certain nuclear reactions. In other cases, modular arithmetic was found of use.

To say this, though, is not to belittle in any way the work of those scientists who, in the closing years of the last century and the first third of the present one, rewrote man's textbooks of the atom and the universe. Around 1900 physicists did not have to wait for new mathematical tools as they had had to wait in the earlier, Newtonian revolution which took place after 1650. And these later scientists were the first to acknowledge how much their achievements would have been delayed but for the work of those mathematicians, pure as well as applied, who had formed new symbolic languages a generation, a century or even longer before men found the use and the need for them.

Epilogue

Asked to name the most notable technical advances of the last decade, most people would probably answer 'Computers and space travel'. If we have mentioned neither yet it is because a final, brief glance at these subjects will give us a chance to stress two of the themes which have run through this book – the underlying unity of the mathematics of all the ages, and the impossibility of predicting which lines of thought will turn out to be 'useful'.

In the design of space craft themselves, for instance, branches of pure mathematics such as topology and the theory of numbers may have been enrolled to help with problems of wiring and instrumentation. But the rocket can only reach its target through man's knowledge of ballistics, of the flight of a body through space as this is affected by the gravitational attraction of other bodies; and here the line of development reaches back through Newton and Galileo to the Alexandrian analysis of conic sections. The first age of ballistics proper extended from about 1500 to 1700, the period in which ordinary firearms became reliable and when it was realised that the problems of the flights of projectiles were similar to those of the movements of heavenly bodies. Then, after 250 years, more complex questions of the same type came to be examined as a result of the development of rocket-propulsion; and this re-examination itself had the twin aspects of application to warfare and of bringing men into a closer contact than previously with the depths of space.

It was largely the difficulty of solving the systems of simultaneous equations involved in modern ballistics – at first of reaching results in a few seconds when training anti-aircraft guns, but soon afterwards of working in split seconds when making 'in-flight' corrections to the rockets of war and peace – which led to a revival of interest in the type of computing machine which we last saw breaking the heart of Charles Babbage. The problem which had defeated him was the same as had given Pascal such difficulties in constructing his simple adding machine, the problem of friction. And since this same force hindered the development of 'thinking machines' using even electro-magnetic relays, computers did not become practical until the present electronic age of first the valve and then the transistor.

Something has been said about the 'hardware' of computers in another book in this series, however; and what concerns us now is their 'software' – the way in which they are mathematically 'programmed' to deal both with calculations on paper (such as the compiling of tables of the type which Babbage himself worked on) and with the 'cybernetic' or 'real-time' problems of the direct control of factory machinery or a rocket in flight. In particular, we are interested here in one aspect of the method of working of the type of computer developed from the adding-machine line of Pascal and Babbage and known as 'digital'. The other branch of the family – comprising the 'analogue' computers whose own ancestor is the slide-rule associated with Napier's colleagues – has less general interest, though some of its members make very ingenious use of specialised mathematics.

This aspect, which in itself forms the main bridge between the hardware and software of computers, concerns the type of arithmetic used; for this takes us right back to Babylonian times and to that idea of a 'radix' which lies at the heart of numbering. We have seen how, over the centuries, 10 became accepted as the radix for almost all arithmetic; but from time to time mathematicians had experimented with alternative roots, and in particular they had discovered that a number-scheme with the simplest radix of all, 2, had some intriguing properties. For instance, this scale (in which the integers 1, 2, 3, 4, 5, etc..., appear as 1, 10, 11, 100, 101 and so on, with an extra figure-place being needed every time we meet a new power of 2) gave the ancient Indian mathematicians a handy short-cut answer to some of their brain-teasers. And centuries later it appealed to the mystical Leibniz, who saw in it an illustration of how God could create *everything* out of only nothingness (0) and Himself (1).

More practically, an English mathematician working about fifty years ago suggested that this 'binary' arithmetic would provide the best possible language for mechanical calculating machines because its two elements – the unit and the zero – were as simple as something happening and nothing happening. In electrical terms this meant that a current either flowed or did not flow. But more than another generation had to pass before the electronic circuitry was ready by which this idea was translated into 'black boxes' which could combine and shunt around chains of pulses of electricity, themselves representing numbers, without translating these into mechanical movements before the final 'print-out' stage. After 1950, however, developments came unexpectedly fast, so that today computers are familiar inhabitants of the worlds of commerce, industry and government as well as of pure and applied science.

The fact that digital computing is based on the two electrical words 'Yes' and 'No' leads to an important parallel: for the symbolic algebras of Poole and Russell largely enshrine the language of classical logic ('No cats are dogs') and hence the *affirmations* and *negations* which in turn can be combined into more complex ideas such as *alternatives* and *contingencies*. And so, as Babbage had foreseen, a versatile calculating-machine had really to be a logic-machine, the highly abstract work of Lewis Carroll's contemporaries acquired a very practical interest, and men were guided to the moon by the kind of circuitry which was most at home when playing a 'Nim' game of picking-up matchsticks. In modern technology the identification between computer electronics and computer logic is so complete that the final pin-point-sized bits of switchgear from which circuits are sewn together are essentially devices for saying 'And', 'Not', 'If', 'Or' and the like.

Actual mathematics, in fact, is not the strongest point of ordinary digital computers, and so these are protected even in their special 'memory' and 'arithmetic' units from the need to solve any sums much harder than $1+1 =$ (binary) 10. The fact that they are nevertheless used to carry out calculations far beyond the limit of human patience (though not, of course, *power*) is itself due to three facts. The first of these is that they can be used in conjunction with the specialised analogue computers which can always be designed to handle such complex but standardised problems as solving the systems of differential equations involved in (for instance) weather-prediction. The second is the prodigious speed of electronic 'thinking': millions of times faster than that of the human brain, computers can afford to use clumsy trial-and-error methods when applying discrete processes to problems in the continuous.

And, finally, a host of problems in equation-solving can be translated from algebraic or even 'analytical' terms into those of the computer's digit-counting by the use of the tricks of the trade which we have called algorisms. The introduction of the computer, indeed, has revived interest in these algorisms amongst some mathematicians – and earned the contempt of others. For there will always be those who love a game for its elegance and skill and those who play only to win – and the conflict of the two schools is itself a fertile one.

The development of computer mathematics largely took place in the United States and – like so much modern technology – depended on team work. One or two names however, stand out, such as those of Warren Weaver and Norbert Wiener, the second of whom studied under both Bertrand Russell and David Hilbert. Nor could any mention of mathematics in the present century omit mention of Jon von Neumann, a former infant prodigy who emigrated from Hungary to America and became a pioneer of atomic power, the theory of games and rocket-ballistics as well as of computer mathematics. Von Neumann, in fact, is the twentieth century's leading representative in that line of mathematicians which perhaps began with Archimedes and which certainly runs through many of the great men of the renaissance and Napoleonic periods down to Gauss and Kelvin.

These were men who particularly interested themselves in the social applications of mathematics, and who were hence often looked to by their governments as professional problem-solvers. If our technological civilisation is to endure it will have increasing need for such men. But if it is to continue breaking new ground, with every generation witnessing achievements which the previous generation would have regarded as almost miraculous, then it will owe as much to the other tradition of mathematics – that which began with Thales and Pythagoras and which is today represented by the specialists in such 'useless' studies as topology and logic, by those working on the theories of sets and numbers, and possibly even by the philosophers who are still wrestling with the questions of the nature of mathematics and of its parts and processes.

Awaiting any of these men – or their successors working in branches of the subject as yet unchristened, for in mathematics as in science progress comes by a continual regrouping of forces – may be the discovery of a tool as powerful and versatile as the infinitesimal calculus or the theory of probability. But just as these were first developed largely because men were interested in curves and chances for their own sake, so the practical advances of the future will be rooted in discoveries which were made out of pure curiosity, intellectual adventure and even a sense of beauty – the beauty which informs some of the simplest as well as some of the most complex mathematics, the beauty which, like mathematics itself, is in the words of St Augustine 'Ever ancient – and ever new'.

Appendix

This book, like others in the series, ends with a table of the men whose lives have been mentioned in it. The conventions previously used have been followed in general, but the grouping has been accommodated to the fact that the growth of mathematics was rather steadier than that of the physical sciences. In principle the time-brackets correspond to chapters, and the men included under each heading are those whose creative lives were mainly spent in that period and whose birth-dates accordingly fall at least twenty-five years before its close. A number of mathematicians have been mentioned in the text out of their strict chronological order, however, and this is noted in the summary by a reference to the relevant chapter.

As usual an * denotes that the man referred to carried out most of his work away from the land of his birth and/or formal nationality. Modern politico-geographical terms are used, and the terms 'Greek' and 'Alexandrian' include the whole of the ancient Greek empire. Both here and in the text, liberties have been taken in such matters as the transliteration of Russian names.

Before 1000 BC (*Chapter 1*)

		BC	
Ahmes	fl. *c.*	1700	Egyptian

999–1 BC (*Chapter 2*)

Thales of Miletus	?	624– 546	Greek
Pythagoras	*c.*	569– 500	Greek

Archytas	fl. *c.*	400	Greek
Eudoxus of Cnidus		408– 355	Greek
Euclid	*c.*	330– 275	Greek
Archimedes		287– 212	Alexandrian Greek
Eratosthenes	*c.*	275– 195	Alexandrian Greek
Apollonius of Perga	*c.*	262– 200	Alexandrian Greek
Hipparchus	fl. *c.*	175	Alexandrian Greek

AD *1–1580* (*Chapter 3*)

		AD	
Hero of Alexandria	fl. *c.*	50	Egyptian
Diophantus	fl. *c.*	260	Alexandrian Greek
Pappus	fl. *c.*	300	Alexandrian Greek
Brahmagupta	fl. *c.*	628	Indian
Hovarezmi M.b.M. al Khovarismi	fl. *c.*	825	Arab
Gerbert (*Pope* Sylvester II)	*c.*	950–1003	French
Omar Khyyàm	?	1044–1123	Persian
Gebir-ben-Aflah	c.	1080–1145	Arab
Abraham ben Ezra (*Rabbi*)	?	1095–1167	Jewish
Leonardo of Pisa (Fibonacci)	?	1175–1250	Italian
Oresme, Nicole (*Bishop*)		1323–1382	French
Müller, Johann (Regiomontanus)		1436–1476	German
Ferro, Scipio		1465–1526	Italian
Niccolo Fontana (Tartaglia)	?	1500–1557	Italian
Joachim, George (Rheticus)		1514–1576	Austrian
Viète, François (Vieta)		1540–1603	French
Stevin, Simon (Stevinus)		1548–1620	Belgian
Napier, John (*Lord*) (Chapter 4)		1550–1617	Scottish

1581–1668 (*Chapter 4*)

Harriot, Thomas (Chapter 3)	1560–1621	English
Briggs, Henry	1561–1630	English

Name		Dates	Nationality
Mersenne, Martin (*Fr:*)		1588–1648	French
Desargues, Gerard (Chapter 5)		1593–1662	French
Descartes, René		1596–1650	French *
Cavalieri, Bonaventura		1598–1647	Italian
Fermat, Pierre (Chapter 5)	?	1601–1665	French
Wallis, John		1616–1703	English
Kaufmann, Nicolaus (Mercator)	*c.*	1620–1687	Danish *
Pascal, Blaise (Chapter 5)		1623–1662	French
Huygens, Christiaan		1629–1695	Dutch
Barrow, Isaac		1630–1677	English
Gregory, James (Chapter 5)		1638–1675	Scottish
Newton, Isaac (*Sir*)		1642–1727	English
Leibniz, Gottfried W. (*Baron*)		1646–1716	German

1681–1800 (*Chapter 5*)

Name		Dates	Nationality
Bernoulli, James (I)		1654–1705	Swiss
Bernoulli, John (I)		1667–1748	Swiss
de Moivre, Abraham		1677–1754	French *
Taylor, Brook (Chapter 4)		1685–1741	English
Maclaurin, Colin (Chapter 4)		1698–1746	Scottish
Euler, Leonard		1707–1783	Swiss *
d'Alembert, Jean-Baptiste le R.		1717–1783	French
Lagrange, Joseph I (*Count*)		1736–1813	French
Monge, Gaspard (*Count*)		1746–1818	French
Laplace, Pierre-Simon (*Marquis*)		1749–1827	French
Legendre, Adrien-Marie		1752–1833	French
Fourier, J–B. Joseph (*Baron*)		1768–1830	French

1801–1875 (*Chapter 6*)

Name		Dates	Nationality
Gauss, Karl F.		1777–1855	German
Bessel, Friedrich W.		1784–1846	German
Poisson, Simeone D.		1781–1840	French

Poncelet, Jean-Victor (Chapter 5)	1788–1867	French
Cauchy, Augustin-Louis (*Baron*)	1789–1857	French
Möbius, August F.	1790–1863	German
Babbage, Charles (Chapter 7)	1792–1871	English
Lobachevsky, Nikolai I	1793–1856	Russian
Abel, Niels H.	1802–1829	Norwegian
Bolyai, Janòs	1802–1860	Hungarian
Jacobi, Karl C. J.	1804–1851	German
Dirichet, Peter G. L.	1805–1859	German
Hamilton, William R. (*Sir*)	1805–1865	Irish
De Morgan, Augustus (Chapter 7)	1806–1871	English
Kummer, Ernest E.	1810–1893	German
Galois, Evaniste	1811–1832	French
Sylvester, James J.	1815–1897	English
Boole, George (Chapter 7)	1815–1864	English
Weierstrass, Karl W.	1815–1897	German
Cayley, Arthur	1821–1895	English
Hermite, Charles	1822–1902	French
Kronecker, Leopold (Chapter 7)	1823–1891	German
Riemann, G. F. Bernhard (Chapter 7)	1826–1866	German
Dedekind, J. W. Richard (Chapter 7)	1831–1916	German
Maxwell, J. Clerk	1831–1879	Scottish
Dodgson, Charles L. (*Rev.*) ('Lewis Carroll')	1832–1898	English
Mach, Ernst	1838–1916	Czechoslovakian
Lie, Sophus F. L. P. (Chapter 7)	1842–1899	Norwegian
Cantor, Georg (Chapter 7)	1845–1918	German
Klein, Felix	1849–1925	German

After 1875 (Chapter 7)

Lorentz, Hendrick A.	1853–1928	Dutch
Poincaré, Henri (Chapter 6)	1854–1912	French
Peano, Guiseppe	1858–1932	Italian
Hilbert, David	1862–1943	German
Minkovsky, Hermann	1864–1909	Russian
Russell, Bertrand (*Lord*)	1872–1970	English
Einstein, Albert	1879–1955	German *
Keynes, J. Maynard (*Lord*)	1883–1946	English
Ramanujan, Srinivasa	1887–1920	Indian *
Fisher, Ronald (*Sir*)	1890–1962	English
Weaver, Warren (Epilogue)	1894–	USA
Wiener, Norbert (Epilogue)	1894–1964	USA
Dirac, Paul	1902–	English
von Neumann, Jon (Epilogue)	1903–1957	Hungarian *

Indexes

The index of this book, as of others in the series, has been divided into two parts. The *General* section contains references to persons, places and the more important of general themes. In the *Technical* index will be found all references to mathematical terms, etc., so that this serves also as a glossary. In both sections, allied words such as Arab, Araby, Arabic and Arabia have been grouped together under one stem for simplicity and to avoid cross-referencing.

GENERAL INDEX

TECHNICAL INDEX